AF651829

[illegible]

[illegible] still exist [illegible] Mongsangei [illegible] boundary of Imphal Municipality.

[illegible] mode of transport for [illegible] products, [illegible] were carried [illegible] Burma and China. Inside the kingdom [illegible] 1820 [illegible] sixteen [illegible] Assam [illegible] In about 1640, King Khagemba improved roads for better transportation of goods and people. King Paikhomba improved road from Chinga to Mongsangei in 1073 (Singh, 19[illegible]5). He

LOCATION, SPACE AND DEVELOPMENT

About the Editors

Konthoujam Indrakumar teaches Political Science in Tamenglong College, Manipur. He is currently working on development and security issues in India and South East Asia. He is Editor, *Alternative Perspectives*.

Homen Thangjam received his doctoral degree from Jawaharlal Nehru University, New Delhi. He specialises in democratization in post-communist and po[illegible]-colonial societies. He is a keen observer of political development in the North East India and his area of interest includes political economy and developmental politics. He is lecturer in political science at the Maharaja Bodhchandra College, Imphal.

LOCATION, SPACE AND DEVELOPMENT

Development Trajectories in India's North East

Edited by
Konthoujam Indrakumar
Homen Thangjam

Published for

THE CENTRE FOR ALTERNATIVE DISCOURSE, MANIPUR

by

CONCEPT PUBLISHING COMPANY PVT. LTD.
NEW DELHI-110059

© Centre for Alternative Discourse, Manipur

All rights reserved. No part of this work may be reproduced, stored, adapted, or transmitted in any form or by any means, electronic, mechanical, photocopying, micro-filming recording or otherwise, or translated in any language, without the prior written permission of the copyright owner and the publisher. The book is sold subject to the condition that it shall not, by way of trade or otherwise, be lent, resold, hired out, or otherwise circulated without the prior publisher's written consent in any form of binding or cover other than that in which it is published.

The views and opinions expressed in this book are author(s) own and the facts reported by him/her have been verified to the extent possible, and the publishers are not in anyway liable for the same.

ISBN-13: 978-81-8069-728-9

First Published 2011

Published and Printed by

Concept
Since 1974

Concept Publishing Company Pvt. Ltd.
Regd. Office :
A/15-16, Commercial Block, Mohan Garden
New Delhi-110059 (India)
Phones : 25351460, 25351794, *Fax* : 091-11-25357109
Email : publishing@conceptpub.com
Website: www.conceptpub.com

Editorial Office :
H-13, Bali Nagar, New Delhi-110 015, India.

Cataloging in Publication Data--*Courtesy:* D.K. Agencies (P) Ltd. <docinfo@dkagencies.com>

Location, space and development : development trajectories in India's north east / edited by Konthoujam Indrakumar, Homen Thangjam.
p. cm.
Contributed articles; with special reference to Manipur, India.
Includes bibliographical references and index.
ISBN 9788180697289

1. Economic development--India, Northeastern. 2. Economic development--India--Manipur. 3. Industrial location--India, Northeastern. 4. Industrial location--India--Manipur. 5. Space in economics. 6. India, Northeastern--Economic policy. 7. Manipur (India)--Economic policy. I. Indrakumar, Konthoujam. II. Thangjam, Homen.

DDC 338.9541 22

Preface

The present edited volume is an overall product of a series of articles published thematically in the second two issues of *Alternative Perspectives*, a quarterly journal. The issues attempted to critically intervene into the debates surrounding the development trajectory of the North East region of India in general and Manipur in particular. Principal reasons cited for underdevelopment of the region in many of the development discourses are that it is landlocked and geographically occupies a formidable mountainous terrain and that it is linked to mainland India *via* a thin 'chicken neck' (Siliguri corridor which is approximately 22 km. wide). Being so, development efforts are arduous and nearly impossible. Apart from initiating a sustained academic engagement on the said issues, as a challenge, drawing comparisons from similar geographical spaces of the world, the editions also aim at providing critical and constructive evaluation of the existing conditions and policies that are detrimental to development of the region.

The *Alternative Perspectives* aims to provide critical and constructive evaluation of the existing conditions and circumstances surrounding the North East. It also offers informed debates on issues crucial to the material realities in the region. The *Alternative Perspectives* represents the recognition of the historical necessity of accepting the twin challenges of to critique and to explore various alternatives for ushering in a new order. The journal is published by the CADM (Centre for Alternative Discourse, Manipur), which is run by Research Foundation Charitable Trust, a registered body, constituted under the Indian Trust Act, 1882, and registered under Registration of Societies Act, 1908. The Centre for Alternative Discourse, Manipur is an outcome of year-long discussion and consultation initiated by the academicians, intellectuals, scholars, thinkers and social activists of the state with the objective to promote a sensible, systematic and informed study on various issues confronting the North East Region through the publication of an academic journal. The very inception of the idea of establishing the centre came from the recognition of a need to critique the realities surrounding our social existence. The establishment of the Centre for Alternative Discourse, Manipur in September 2005 represents our understanding that we belong to a critical historical epoch. An acutely felt need to engage the emerging as well as persisting realities is crucial behind the genesis of the centre.

The present book is the second in a series of publications by CADM. Taking due cognizance of the widespread appreciation and acknowledgement that we received from the academic community across the country for our humble initiatives, we at the CADM, deeply felt the need to produce the second volume of the published works in book form. The present publication has also been necessitated by the reality of underdevelopment in the region, which is mainly due to misplaced difficulties given to the geographical space and misdirected policies ensuing from such a fallacy by policy formulators in New Delhi as well as in various capitals of the States in the region. The result of such policies on the region needs to be properly scrutinized *vis-à-vis* the growing aspirations of the peoples of North East.

We look forward to positive criticism and suggestions from readers for further improvement of the present volume.

Acknowledgements

The Centre for Alternative Discourse, Manipur acknowledges the help and assistance extended by members in finalising this volume in the shape of a book. Foremost, we thank our esteemed readers, for their constant support and criticisms, which has helped us in improving ourselves. We thank Moirangthem Prakash, (Asst. Professor, IRMA, Anand) and Sibam Haripriya (Research Scholar, University of Delhi) for editing some sections of the manuscript. We also thank Laishram Nandalal (Sub Editor, *The Pioneer*, New Delhi), Hidam (O) RK Smejita and Hidam Premananda (Research Scholar, University of Delhi), for their constant support and inputs in terms of knowledge and material.

The Centre misses the presence of Late Thingnam Kishan, Founder Editor, *Alternative Perspectives* (2005–mid 2008), who was instrumental in giving shape to the journal as well as guided to its success. The collection of essays in the present volume was published in the journal during his editorial tenure.

Finally, we would like to thank Concept Publishing Company Pvt. Ltd., New Delhi for publishing this book.

Contents

Contributors

Arambam Noni teaches Political Science in Moirang College, Manipur. The author extensively works on issues related to political theory & political history and conflicts, concerning the North East India.

Hanjabam Isworchandra is a ICSSR doctoral fellow at the Department of Economics, Manipur University, Canchipur. He is working on 'Rural Non-Farm Sectors of Manipur'for Ph.D. programme.

Homen Thangjam received his doctoral degree from Jawaharlal Nehru University, New Delhi. He specialises in democratization in post-communist and post-colonial societies. He is a keen observer of political development in the North East India and his area of interest includes political economy and developmental politics. He is Lecturer in Political Science at the Maharaja Bodhchandra College, Imphal.

Kangujam Sanatomba is a doctoral fellow in Department of Political Science, Manipur University, Canchipur. His doctoral work is on 'Comparative Insurgencies'. He has contributed extensively on insurgency and ethnic conflict for the English daily *The Sangai Express*.

Konthoujam Indrakumar teaches Political Science in Tamenglong College, Manipur. He is currently working on development and security issues in India and South East Asia. He is Editor, Alternative Perspectives.

Laishram Churchill an environmentalist who takes keen interest in social and development issues of Manipur.

Roji Rajkumari completed her M.Phil. from the Department of Political Science, University of Delhi. She specialises in politics relating to Centre-State relations.

Late Thingnam Kishan, former Editor, *Alternative Perspectives,* is author of *Rethinking Colonialism* (Delhi & Kolkata : World View Publication, 2006) and *Look East Policy & India's North East* (New Delhi:Concept Publisher,2009). He was former Assistant Professor in the Department of English, Manipur University.

Thingnam Sanjeev, an archivist by training and earlier Research Assistant with the Oral History Division, Nehru Memorial Museum Library, New Delhi is currently working on Military History at the Jamia Milia Islamia University, Delhi.

Yambem Imo Singh teaches Social Geography at D.M. College of Arts, Imphal. He had worked on Aspects of Surface Transport in Manipur for his doctoral thesis.

Introduction

Konthoujam Indrakumar & Homen Thangjam

Economies of the world are interconnected in one way or the other. No single country can remain economically isolated. The neo-liberal wave sweeping across the world under the garb of globalization has made it more evident; today, the world is considered to be a global village. Advance in science and technology and revolution witnessed in the information technology belies geographical isolation. Landlockedness is no longer considered an impediment to a country's development or economic growth. However, mere integration into a larger economy or the global economy does not guarantee development of the landlocked country or region. A larger polemics as witnessed in many parts of the developing countries are examples to ever growing instances of exploitation and economic subjugation by a richer or stronger country. From this perspective, scholars in different parts of the world have condemned the integrationist mode of globalization especially in the realm of trade and commerce as extension of colonialism and popularly term the wave as neo-colonialism which destructs economy, culture and polity of less developed or developing by the developed countries. Chief argument is that mere integration without 'developing' basic economic structures and forcing the weaker nations to accept terms of trade and commerce make such nations dependent on the stronger for durable and other finished goods. In fact, it destroys the productive base of the dependent nation, a situation of 'trap' from which it can never escape. Such arguments carry notes of realism and inform the underdeveloped condition of the North East region of India.

North East region's colonial encounter (read British colonialism) (in different chronological order) marked introduction/integration of the region to global capitalism / imperialism. Shift from feudal mode of production to capitalism did not necessarily mean emergence of native bourgeoisie or a working class, as establishment of factories and other manufacturing units were conspicuously absent. Inherent logic of imperial integration was extraction of surplus from any given source, viz. forest, agriculture or others. Moreover, in the eyes of the colonial power, the region was regarded as an 'enchanted frontier' and an essential security zone towards its extension or preservation of power in the South East Asia corridor. Thus, conception of the region as a frontier nullified any efforts for development in the region.

Notion of numerous variables regarded as barrier to development in the region continued to be 'preserved' by the Indian State in the post-Independent era. For example, the region was termed as landlocked inhabiting a hostile geographical terrain, and that such are the impediments to development. Reflection of such misconceived ideas is found in the policy formulations for development in the region. To some extent, it is true that the present geography of North East possesses characters of landlockedness; it is enclosed by land on all sides and thus suffers from sea trade or access to sea port. However, such a proposition is devoid of historical truth as the present map of North East is a creation mainly driven by 'partition' of Bengal in the west, appeasement policy to Myanmar and China in the South East and East Asia, respectively. As a result, historical trade routes and access to sea across the Chittagong Hill Tracts (for example Cox Bazaar) in present day Bangladesh was severed. On the other hand, access to Chindwin River that flows into the Bay of Bengal was denied.

The artificial contrivance mainly pertaining to landlockedness of North East region was debunked by policy makers in New Delhi, when they proposed a deep sea port (DSP) at Cox's Bazaar district of Greater Chittagong. Chittagong port is located at only 75 km. from South Tripura township of Sabrum. The proposed port was mainly to facilitate movement of transhipment cargo to North-Eastern India. Currently shipping lines that ferry cargo to Bangladesh have to unload at Haldia or Kolkata port. The cargo is then transported to the neighbouring country *via* road. With another port in the neighbourhood, shipping lines would have an economically viable option in terms of transportation of their goods. Bangladesh is almost at the centre of South East Asia, the DSP could act as a regional hub to transport goods to North East India. Apart from improving movement of transhipment cargo in the North-Eastern part of India, the DSP could also prove useful to other neighbouring nations like Myanmar, Southern China, and landlocked countries like Nepal and Bhutan.

However, as Bangladesh was reluctant to co-operate the DSP venture, on the ground that the Chittagong port was unable to cater to the domestic demands of Bangladesh, New Delhi proposed to open a new route for the region through Myanmar avoiding Bangladesh. The proposed port is at Akyab (Sittwe) port on the Bay of Bengal in Myanmar as a transit seaport for a sea-and-river trade route for the North East region through Mizoram. The new plan envisages goods transportation from Akyab along the river Kaladan to Mizoram. The distance between Kaladan to Mizoram is less than 200 km. Opening an alternative new route for the North East ships can also sail to Kolkata or Visakhapatnam from Akyab.

What has shaped India's imagination of the North East, from landlocked to 'potential area of opportunity' (after nearly six decades) is its eagerness to go ahead with its Look East policy. The current official understanding is, 'the North East is region rich in resources. However, due to lack of access, the potentials have always been underutilized. A gateway through Burma will allow a resource-led growth in the North East' (Dipankar Chatterjee, Chairman of the Confederation of Indian Industry's North-Eastern Council quoted in *The Indo-Burma News*, April 24, 2009). But the question remains, if the Policy (read Look East Policy) stands to benefit the people of the region. For more than five decades after the departure of the British, if India has made North East region a dependent region on finished goods from mainland India, without developing the productive bases, the Look East Policy cannot be looked upon without suspicion, as such an extension and integration into a larger economic bloc suggest to convert the region into a transit route and highway economy.

Historically, being landlocked was regarded as a disadvantageous position. It cuts the country off from sea resources such as fishing, but more importantly cuts off access to seaborne trade which, even today, makes up a large percentage of international trade. Coastal regions tended to be wealthier and more heavily populated than inland ones. Thus, many countries made efforts to avoid being landlocked. For example, the International Congo Society, which owned the modern-day Democratic Republic of the Congo, was given a thin piece of land cutting through Angola to connect it to the sea by the Conference of Berlin in 1885. After First World War, in the Treaty of Versailles, a part of Germany, designated 'the Polish corridor', was given to the new post-First World War country Second Polish Republic, for access to the Baltic Sea, which was also the pretext for making Danzig (now Gdansk) with its harbour the Free City of Danzig. This made Poland a semi-landlocked country as described in the previous section, but Poland soon enlarged the small fisher harbour of Gdynia into a large one. Coming to recent events, Serbia became landlocked when Montenegro split from Serbia. However, the country has access to the Black Sea through agreement with neighbouring countries.

Losing access to the sea is generally a great blow to a nation, politically, militarily, and particularly with respect to international trade and, therefore, economic security. For example, the creation of the new states of Eritrea and Montenegro, brought about by successful separatist movements, has caused Ethiopia and Serbia respectively to become landlocked. When the Entente Powers divided the former Ottoman Empire under the Treaty of Sèvres at the close of First World War, Armenia was promised part of the Trebizond Vilayet (roughly corresponding to the modern Trabzon and Rize provinces in Turkey). This would have granted Armenia access to the Black Sea. However, the Sèvres treaty collapsed with the Turkish War of Independence and was superseded by the Treaty of Lausanne which firmly established Turkish rule over the area. Currently, the United Nations Convention on the Law of the Sea gives a landlocked country a right of access to and from the sea, without taxation of traffic through transit states. The United Nations has a programme of action to assist landlocked developing countries.

However, the irony is that no concerted effort was undertaken by the New Delhi, barring recent development, which is chiefly driven by India's better economic and security vision in the emerging changed geopolitical configuration in the post-cold war period. Development as an issue remains central in the context of the North East ever since the integration of the region into the world capitalist system through the insinuation of the British colonial policy. It has been conspicuously marked by underdevelopment. With British departure from the continent and the region's integration with the successor Indian State, expectations of the paradigm shift in the development framework have been deluded with massive poverty and underdevelopment still persisting. Development debate on the North East continued to be perceived as complex and inconclusive in mainstream Indian economic discourse. Explanations of the region's underdevelopment from different vantage positions underscored the geographical location, topography, cultural traits of the indigenous population, scarcity of resources and skilled manpower among a horde of other factors as impediments to development. Policies and descriptions have not been able to successfully transform economic landscape of the region.

A polemically charged concept, development continues to remain inconclusive in terms of the inability to define and confine its semantics. Broad based understanding that has emerged over the year's associate development with improvement of living conditions of human population. Economic, political and social conditions are necessarily included as inherent and intrinsic parameters in evaluating development. Debates and discourses on the multi-semantic nature of development underscore the importance and centrality of these aspects of human life. In essence, it connotes the inherent human tendency for desirable changes in society. Human history is clear evidence of this pervasive tendency for change aimed at the betterment of society. Scientific understanding of this tendency increasingly attaches development with economic betterment of society which forms or constitutes the basis of transformation of living conditions in accordance with human aspirations. Continuing the efforts to grapple with the theme of development in the North East perceived as a jinx in mainstream Indian development discourse.

Behind this backdrop the present edition takes a look at the central debates surrounding the issue of development in the North East in an attempt to evolve a critique of development models, and policy prescription witnessed so far. In evolving such a critique the implications of policies rooted in bureaucratic planning from top instead of resource based planning from the ground assume critical importance. Assessment and analysis of notions and perceptions on development in the region form an important part of this attempt. The need to seek reasons behind failure of development policies has propelled this attempt to uncover some of the uncomfortable truths and realities surrounding development debates *vis-à-vis* the region. Moreover, the essays in the collection seek to understand, the dynamics of policy prescriptions in the region.

The first essay by Laishram Churchill titled 'Location and Implications on Development : A Critique on the North East', raises the fundamental question surrounding location as an important issue in development debates. Location of a region or a country in terms of being landlocked is critically explored to understand its relationship with development. The essay provides a comprehensive overview of the theoretical polemics on landlockedness and development. Laishram critiques the ideological assumption behind the notion that landlockedness is a natural corollary to underdevelopment. Specific case studies of certain landlocked regions form the base of his assertion that development necessarily does not depend entirely on the locational disadvantage of being landlocked. The essay offers different means and ways of overcoming locational disadvantage.

The next essay by Konthoujam Indrakumar titled 'Geo-Spatial Reality and Development Paradigm : A Case of the North East Region' aims at delinking the correlation between the topographical features and economic underdevelopment of the region. In it he analyses the cases of Yunan province of China and Macedonia. Yunan as well as Macedonia have striking resemblance with India's North East Region in terms of landlockedness, topographical features, ethnic composition and political instability, etc. that surround the region. However, as revealed in the essay both Yunan and Macedonia have successfully overcome various difficulties posed by geographical features and unrest embedded in the multiethnic component of the province and country respectively with consistent and clear cut development strategy which is conspicuous by its absence in India's North East Region. The author questions the correlation between locational feature of a region and development status based on the study of Yunan province of China and Macedonia, a small landlocked country in the south-western Europe, the author, therefore, argues that lack of development in the region has to be attributed to the policy and programmes initiated by those at the helm of affairs and Centre's perception towards the region which is deeply rooted in the geo-political thinking. In the last part of the essay Konthoujam narrates briefly the pre-British trading pattern, colonial period and post-British phase wherein he argues that post-British period witnesses the consolidation and perpetuation of British colonial policy in the region. Underdevelopment in the region is basically a result of flawed process of development strategy arising out of predominant perception of the region as a military buffer. The crux point of the essay is that development discourse in the region should go beyond locational features and insurgency centric view.

The third essay by Thingnam Sanjeev titled, 'Relocating the North East in Global Economic Space : Dynamics of Periphery' surveys the historical transformation of the region into a frontier periphery subsequent to British colonial intervention. British policy of maintaining the region as a militarily secured 'frontier' underscored the cruciality of location and space in colonial politics. Arguing that traditional trading routes were severely affected by colonial intervention, the essay critiques the 'frontier' legacy in the post-British period. Disorientation in the region's economic configuration is largely rooted in the perpetuation of this legacy. Thingnam underlines the recent moves by the Government of India to restore historical links of the region as paradigm shift that needs to be fully examined.

The next essay in the collection by Arambam Noni Meetei titled 'In the Confines of a Geo-Spatial and Extra-Economic Identity : Development of North East India' traces the uneasy discomforts of a perceived 'otherness' of the North East region in an attempt to demystify the notion of an inclusive development framework enumerated by policy makers and planners in mainstream India development discourse. History of transition from British colonial politics to post-British integrationist politics narrates the prejudiced legacy of mainstream perception of the region which ironically subserved the logic and practice of underdevelopment of the North East. Confinement of the region in terms of a perceived geo-spatiality underlines the inability of policy makers in New Delhi to search for substantial means and ways to generate development. Location, history and politics have combined to take the regions development issue beyond tangible economic rationale. Arambam contends that economic centrality in development planning was altogether conspicuously absent *vis-à-vis* the North-Eastern region since inception of economic planning in post-British India. In what has been termed as extra-economics the essay underscores the geo-spatial regimentation of the region responsible for the warped outlook towards the region.

The following essay in the collection by Sanatomba Kangujam titled 'State and Development in the North East' offers a theoretical critique on the issue of development in the North East *vis-à-vis* the State and its prescriptions. Taking note of the polemical debates surrounding the notion of development, Kangujam obliquely rendered visible in propounding an explicit *raison d'etre* fail to address people's aspirations in the region. Asserting the inextricably intertwined logic of aspirations and development, the essay underlines the statist attempt to ignore aspirations while adopting a prescriptive approach as an important factor behind underdevelopment of the region. Reductive analysis provided by the frame of financial and other relations between Centre and States cannot offer a tangible and substantive understanding of the complex realities in the region. Kangujam's argument calls for a rethinking on the failure to address the complex politico-historical realities of the body politic of the region in order to evolve a holistic understanding of development issues in the region.

The sixth essay by Roji Rajkumari titled, 'Understanding the Politics of Underdevelopment in the North East' provides a critique on the nature of underdevelopment in the region. It surveys the historico-politico-economic and geographical specificities of the region and incompatible response of the Indian State to these specificities along with adverse forms of development (read underdevelopment) in the region. Rajkumari investigates various policies adopted by the Indian Government and failure on its part to come up with appropriate North East policy that can accommodate the people of the region. The essay contends analytical study of the existing conflict and armed struggle in the region and recounts it as deep rooted and enduring in the historical process of the region.

'Economics of Underdevelopment : Reading the North East' by Late Thingnam Kishan Singh, attempts to intervene in the ideological debates on the very nature of development in the North East. It contends that the region's development debates have generated polarized positions. Underdevelopment alongside a systematic plundering of the region's resources form the matrix of dependency during the British colonial period. Post-British policies failed to transform this dynamic as economic development was reduced to the successful establishment of the market for products manufactured elsewhere. Thingnam raises the fundamental issue of production as the basis of meaningful development and argues that development policies have remained detached from production. Until and unless production assumes centrality development will remain lopsided. Meaningful utilization of the region's resources by generating production remains the logical way out of the current impasse.

The next essay in the collection, 'Manipur's Economy from 1949 to 1972 : Birth and Nurturing of a Dependent Economy' by

Hanjabam Isworchandra Sharma traces the transformation of a self-sufficient agrarian economy to a dependent economy through the insinuation of centralised economic planning of New Delhi. On the eve of merger with India in 1949, Manipur's economy had recovered from the setback suffered during the Second World War to retain its position as a surplus exporter of rice, its staple food crop. However, Hanjabam asserts that post-'Merger' transformation of Manipur's political status into a Union Territory left economic planning entirely in the hands of the Central Government. With the Central Government enjoying exclusive prerogative for the first four Five Year Plans before statehood in 1972, anomalous trends in the economic planning significantly eroded the agrarian base of Manipur's economy. Agriculture and allied services were severely neglected along with conspicuous absence of steps towards industrialization. He argues that anomalous planning trends – far removed from pattern followed elsewhere in the country with emphasis on primary and secondary sector – transformed Manipur into a dependent economy. Dependency increased as massive influx of commodities manufactured in the industrialized areas of the country through a trading class of outsiders comprising mainly of *Marwaris* and *Bengalis*.

The last essay by Yambem Imo Singh titled 'Surface Transport and Economic Development in the Manipur State' studies the evolution of surface transport and its tangential links with economic development in Manipur. In a fairly descriptive manner, Yambem provides a detailed account of land routes in Manipur since early times. Identifying land routes as the only major means of transport with marginal presence of water transport in the relatively small rivers, the essay surveys the nature of economic activities supported by surface transport. Trade and commerce in the ancient period carried through narrow and steep roads in the hills and valley involved items like silk, bamboo and cane products, gold and animal horns. Tracing the development of the National Highways 39 and 53 in the post-British period, the essay details the nature of outflows and inflows in Manipur witnessed through these two routes that serve as the most important links with the rest of India. Underlining the crucial aspect of wide disparity seen in the massive inflow of commodities from other parts of the country compared to the meagre and insignificant outflow from the State, Yambem foregrounds the need for substantial production and increased accessibility as twin components of holistic development. The pathetic picture of accessibility as far as the hill region of the State is concerned comes out as a strong point in his assertion that misdirected and haphazard surface transport policies have played with economic development in the State.

1

Location and Implications of Development : A Critique on the North East

Laishram Churchill

Landlocked countries (LLCs) are defined as those countries with no direct coastal access to the sea and thus also to maritime trade. Being landlocked, these countries are deprived of the cheapest means of transport-shipping, putting them in a disadvantageous position as compared to coastal countries. Although economists have accepted that a country's size, location, topography and natural resource endowments influence on prospects for industrialization and growth is still a matter of intense debates. Studies on the impact of landlockedness on the overall development of a country have shown different positions and standpoints. It cannot be denied that landlocked countries are vulnerable on two points. First, such countries are vulnerable on their own account and secondly, on account of being dependent on one or more transit countries. Neighbouring countries enjoy the leverage to extract economic and military incentives to block their access to the sea or transit through their territory. Secondary higher transport costs incurred by whatever infrastructure deficiencies, delays, fees or procedures in the transit country make the landing or shipping of goods from and to landlocked countries very costly and oblige the landlocked country to maintain high level of inventory. In fact, for most landlocked countries, high transport cost is the single most prominent hurdle to their equitable access to global markets and competition with other countries.

Apart from dealing with the problems associated with landlockedness, prospects and opportunities in the context of the modern dynamics of economics, the essay intends to study similar problems faced by hinterland regions of coastal countries with special reference to India and its North-Eastern Region. In fact, the only feature that separates hinterland regions from landlocked countries is the absence of international borders through which export and import business takes place. Latter part of the essay presents certain structural case studies on how some landlocked countries and hinterland regions have managed to overcome certain constraints associated with landlockedness. An important issue behind development debate is the traditional notion of seeking or explaining the causes of economic underperformance of a country or region in 'natural' or geographical factors alone. Gallup *et al.* (1998) suggest that location and climate through their impacts on transportation costs, the common diseases associated with the climate and location of the land and agricultural productivity have significant impact on development. The essay aims to focus on the development problems and prospects of landlocked regions with due consideration of factors such as passage over the territory of transit State and access of landlocked countries to marine resources under International Law of the Sea (Glassner, 1992).

Landlockedness : Theoretical Perspectives

Theoretical underpinning on landlockedness, until recently, was primarily influenced by neoclassical theory. This theory suggests that when goods cross a country, there will be transaction costs having to do with customs and handling. If there is switch in transport modes, there will be off-loading and on-loading costs, and perhaps storage costs as well. Therefore, landlockedness can be thought of as raising the price of imports and reducing the price of exports and further, net transport costs (which in turn must be absorbed by a price taking seller in order to compete in the international market). Thus, landlocked countries (hereafter LLCs) are put in a disadvantageous position when it comes to cross-border trade, particularly overseas, resulting in reduction of real income.

The greater the share of primary commodities in the export basket, and of finished goods in the import basket, the greater the terms-of-trade loss. The income effect of the terms-of-trade loss causes reduction in the volume of imports. Furthermore, the substitution effect is to compel consumers to switch from imported goods to domestic products, reducing the exportable surplus as the opportunity cost of consuming at home instead of exporting falls. Both the import and substitution effects reduce total trade (imports plus exports) in gross domestic product (hereafter GDP). LLCs may import less (for example, engage in inefficient import substitution) and export less (for example, fail to exploit otherwise attractive export opportunities) than would be warranted by comparative advantage exclusive of transport costs.

An alternative interpretation of the theory argues that LLCs pay to the Transit States (hereafter TSs) tariffs on incoming goods and tax on outbound goods. As a transport monopolist, the TS would not kill the goose that lays golden eggs; rather it will charge transport rates that reflect the price elasticity of demand and supply in the LLC. A third interpretation asserts that, when a TS rations the supply of transport to an LLC, the quotas thus created give rise to rents in the form of price gauging and perhaps corruption. Finally vulnerability of the LLC to strategic disruption is a factor often mentioned by some analysts (Mackellar, 2000, 3). However, arguing by analogy from empirical evidence regarding the effect of punitive trade sanctions, the economic impact of trade disruption may sound modest. The uncertainty factor, exacerbated by the often strained political relations between LLCs and TSs, may be more important than actual costs. Indeed, a World Bank study identified reliability of transit, not cost, as the main criterion for project design and appraisal (World Bank, 1994). LLCs access to ports is governed mostly by bilateral treaties with TSs, most of which are fixed for long periods, while many are *ad hoc* and some comprise only a clause or two of larger treaties. Uncertainty over deliveries may make potential customers wary of signing long term export contracts and discourage foreign firms from locating plants and investing in LLCs. However, it is pertinent to note few LLCs are completely dependent on single TS. The dependency relationships that exist can be ameliorated over time by negotiating alternative routes and undertaking integrated regional investment programmes.

While neoclassical theory offers an explanation why LLCs might be at a disadvantage in the short term, it offers little explanation as why this disadvantage might be insurmountable in the longer term. For example, Switzerland despite being landlocked is one of the highly developed countries in the world, with its per capita income being one of the highest. The argument assumes greater significance and rationalism with information technology (hereafter IT) revolution sweeping over the globe, which has substantially reduced space-time constraints and in some cases negated geographical barriers, providing accessibility and connectivity to the international markets through satellites. Furthermore, growth of regional economic blocs or forums such as European Economic Community (EEC) and Association of South East Asian Nations (ASEAN), etc. have considerably mitigated the disadvantages associated with LLCs. In fact, the coming of IT revolution and growth of regional economic forums have shattered the myths associated with landlockedness, thereby creating an inevitability of serious introspection into the credibility of neoclassical theory.

Considering the new opportunities and growth avenues opened to LLCs, the 'new trade policy' has deservedly replaced the 'endogenous growth theory'. Besides investment, human capital accumulation, technological innovation and provision of infrastructure and public goods identified in the 'endogenous growth theory' as sources of growth, the 'new trade theory' incorporates policy measures such as industrial policy, regional policy, trade policy and provision of infrastructure and public support for human capital formation, research and development, etc. (Krugman, 1987). New trade theory rigorously asserts need for the state to play a more proactive role in trade and commerce, and development of necessary infrastructure. Juxtaposition of the neoclassical theory and new trade policy gives little reason to assume that landlocked countries should fare worse than others economically. Even though LLCs are heavily over represented among the poorest countries in the world, this does not necessarily mean that geographical factors or landlockedness are solely responsible for the economic backwardness. For instance, as per 1998 World Development Indicators, out of 20 poorest countries listed, eleven were not landlocked countries. It has been generally accepted that problems associated with landlocked status can be mediated through effective transport policy. At the same time, modern advances made in telecommunications and information technology for development of economic sectors can reduce dependence on physical transport.

Transport Policy

Investment in transport infrastructure is necessary to enhance trading capacity and productivity of any country, coastal or otherwise, but it is of extra significance for landlocked countries. Improvised transport infrastructure leads to a direct reduction in transport costs, by reducing journey and time consumption. Benefits of improvised transport sector have been felt all along generating lower cost for imported capital goods and intermediate inputs, which reduce the bias against certain types of foreign direct investment (hereafter FDI) in landlocked countries. Improvisation of LLCs' infrastructure from the twenty-fifth percentile to the seventy-fifth percentile would eliminate around one quarter of the disadvantage associated with landlockedness (Venables and Limao, 2001). Consequently, improving transport infrastructure should be a policy priority for LLCs, even in the absence of any improvement to the transport infrastructure of the transit country.

In addition, improving infrastructure has the potential to reduce the uncertainty faced by investors in relation to the costs of transportation through the transit country. Increasing the number of export routes and improving their quality makes it less likely (other things being equal) that there will be any disruption to traffic. Improved infrastructure can considerably mitigate the impact of currency volatility by reducing the amount of the transit cost subject to random fluctuations in the bilateral exchange rate between the TS and the LLC.

Regional Integration : A Powerful Tool

Regional integration and economic blocs can be very powerful tools for mitigating the problems of landlocked countries. It has been proven beyond doubt that economic blocs like European Union (EU) and ASEAN have successfully fostered cross-border trade, raising rates of return on investment in transport infrastructure and thus encouraging private sector to invest in transport infrastructure. By lowering transport costs including tariffs and taxes, the resultant improvement in infrastructure not only enhances the terms of the trade of LLCs, but can also substantially reduce the expected cost of transportation in future. As the cost of transportation through a transit country falls, fluctuations in the bilateral exchange rate between the LLC and the TSs become less significant in terms of their potential effect on investment in the landlocked country.

Another major advantage of regional integration is that it can create conditions that lead to reduction in the costs (for example, transhipment, time taken for dispatchment and delivery of goods, bureaucratic obstacle, customs cost and so on) of border crossings between an LLC and TSs. To create such favourable conditions, serious diplomatic efforts should be put towards unification of customs and immigration requirements among the countries in a bloc, introduction of transport permits which are valid in all the countries of the bloc, collaboration and coordination among frontier authorities.

Regional integration can also bring together markets geographically by making it unnecessary for a landlocked country exporter to cross additional borders. Exports from an LLC belonging to an economic bloc would have to cross one border instead of two or more, strictly in terms of customs, taxes and tariffs. As such, regional integration can considerably diminish the competitive disadvantage created in this respect by the remoteness of landlocked countries or regions from international markets.

Effective coordination of multinational infrastructure projects is another benefit of regional integration. Negotiations among countries belonging to a trade bloc can certainly enhance the chances of credit extension and investment from another country or private sector in a given infrastructure project, as the member countries of a bloc are bound by international treaties and agreements thereby allaying the fear about possible change of policy or regime and other political uncertainties in a member country. Thus, regional integration can reduce the effects of coordination failures regarding development of infrastructure projects. Once regional integration succeeds in setting up multinational markets in which firms from different countries do business, several trade routes would be available to an LLC, instead of depending on a single transit country. Reforms in transport regulations of a landlocked country should be guided by two primary objectives such as minimising the impact of regulations on trade and harmonising transport rules with those applied in transit countries, to reduce the cost of business that operates in both the jurisdictions.

According to UNCTAD (2001), international conventions and regional (sub-regional and bilateral) agreements are the main instruments under which harmonization, simplification and standardization of rules and documentation can be achieved. For instance, in South America, the members of Mercosur and the Andean community have undertaken to apply common transit procedures.

Harmonising Documents and Procedures

Trade facilitation procedures have been developed on the national, regional and international levels. Such procedures range from common customs declaration documents to the electronic transmission of data ahead of the arrival of cargo at the tran-shipping point to sophisticated computer programs. Information technology, intermodal transport or other new trends can certainly contribute significantly to improvement of transit traffic and reduction of transport costs for landlocked countries. Simplified systems, which improve the cooperation of customs and authorities in transit and landlocked countries or within one region, are key to reducing transport costs and time.

The EU's Single Administrative Document is one of the

successfully operational simplified systems. The Baltic Common Transit Procedure, which so far covers road transport only and came into force in 2001, is another fine example of simplified systems. The Baltic Common Transit Procedure considerably harmonised and simplified transit through the three Baltic states by introducing a single customs declaration and guarantee. Consequently, the Baltic countries conceived this agreement as a major initiative towards joining the Convention on Common Transit, which is applicable to the transit of goods from/through the EU, EFTA (European Free Trade Agreement) and the Visegrad countries. The Convention of Common Transit and subsequent reforms introduced in 2001 proved to be a major breakthrough of legislative changes, closely inter-related with operational reforms. The basic guiding force of this initiative is to improve the legal environment for transit operations on the one hand specially with regard to avoiding detrimental administrative procedures and fraudulent practice in inter-state trade and on the other, to link more closely the twenty-two customs administrations of the parties to the convention. The need for a common customs document has also been acknowledged in Africa too, particularly within Common Market of Eastern and Southern Africa (COMESA). Similarly, the Sub-Regional Transport Forum of the greater Mekong sub-region has taken initial steps towards harmonization of documents and procedures (Jakob, 2003).

Information Technology

Although investment in IT (Information Technology) sector is a costly venture, it has become an efficient means of reducing detrimental effects of geographical distance and barriers on the trading systems of regional and international levels. Thus, investment in this sector has become important for both transit and landlocked countries. Paper-backed transit systems cause delays and consume extra manpower which is endemic. The EU relied for long time on paper-backed transit procedure but it has recently introduced an IT based documentary exchange system. Now, it has been widely accepted that the paper-backed regime, designed for fewer member states, could simply no longer cope with today's ever growing transit traffic. The huge volume of papers which need to be processed every single day is a huge burden to the customs services and resulted in serious delays in the treatment of transit operations besides causing administrative errors and adverse financial impacts. Several national and international organisations, in their efforts to facilitate trade, are already involved in computerization of procedures. One outstanding example of such a grand project is the Trade and Transport Facilitation in South East Europe Project (TTFSE) of the World Bank, South East European Cooperative Initiative (SECI) and the US (and in collaboration with the EU). The project has brought about physical improvements to border crossings, technical assistance to strengthen customs administrations, computerization of procedures at the border crossings, improved exchange of information between the border control agencies and the business community through seminars, training and internet websites. Membership in the TTFSE Programme is based on being a recipient of funding for customs reform under a loan from the World Bank or a credit from the World Bank's International Development Agency (IDA) and signing a joint Memorandum of Understanding (MoU). By signing the MoU, the participants commit to joining the Regional Steering Committee of TTFSE and to collaborating in the resolution of common problems constraining trade in the region. The MoU includes a direct reference to the improvement of transit and cross-border problems (Jakob, 9). Other examples of concerted effort towards trade facilitation and computerization of procedures are UNCTAD, ASYCUDA and ACIS systems.

Over time, rail, land and air transport as well as telecommunications and IT, have reduced the advantage enjoyed by coastal countries over LLCs. Nonetheless, shipping still remains as the predominant mode of transport for global trade, and as such geographical location of a country remains a crucial factor to its economic growth (Smith, 1976). Although being landlocked is a challenge, it is not a destiny. A brief case study of Hungary would give more insight into how the difficulties and disadvantages associated with LLCs can be effectively addressed.

Hungary : Landlocked in the Heart of Europe

Hungary provides one of the best examples regarding the argument behind landlockedness and development. It amply demonstrates that being landlocked is neither an insurmountable obstacle to the transition process nor to trade expansion or economic well-being. Hungary can access the sea *via* inland waterways, rail and road—with roads being the most important—followed somewhat closely by rail. Inland waterway traffic is by far the least important (Commission of the European Communities, 2001). Trains from Budapest take not more than 36 hours to reach two German ports that have become very important for Hungarian trade to destinations overseas (including the UK and Ireland). The opening of a new railway line between Slovenia and Hungary in May 2001 made the establishment of regular container trains between Budapest and the Port of Koper in the Adriatic Sea possible. According to a November 2001 study, Hungary's railway network ranked fifth among 16 European countries (Healey and Baker, 2001).

The legislative framework which regulates Hungary's transit and access to the sea are codified in bilateral inter-governmental agreements which often also regulate road freight traffic quotas through a system of permits. In July 2000, Hungary and the EU signed the bilateral Road Goods Transit Agreement with the primary objective of facilitating transit across the territory of the contracting parties, particularly through the mutual exchange of road transit authorizations. These authorizations are, over and above, to those already exchanged within the framework of bilateral agreements between the EU member-states and Hungary. In addition, Hungary is also a party to several multilateral agreements including the TIR Convention (Transports Internationaux Routiers/International Road Transit). Customs transit procedures are regulated by the EU Convention on Common Transit of 1987 (amended in July 2001) to which Hungary is a party and is broadly in line with the provisions of the EU Community Transit. The Common Transit regime is not compulsory and the TIR system can be used as an alternative. Hungary also has bilateral agreements on transit and related subjects with other countries, including with those that are crucial for the country's access to the sea. Customs cooperation is also incorporated in the Free Trade Agreements concluded within the framework of CEFTA (Bulgaria, Czech Republic, Poland, Romania, Slovakia and Slovenia), as well as with Croatia and others. The TIR Convention, which is currently used by more than 32,000 transport companies in over 50 countries in Europe, Central Asia and the Middle East, allows road transport operators to cross borders in international and transit traffic without involving major procedures and costs.

Apart from the multilateral conventions, Hungary has also signed two bilateral agreements, one with Germany (1989) and the other with the Netherlands (1991) just prior to the opening of the Rhine-Main-Danube canal in 1992. The two agreements cover the reciprocal use of inland waterways by vessels for the carriage of goods and passengers and for transit and the equal participation of these countries in their respective bilateral trade. Ships of these countries are guaranteed the same treatment while on the territory of the other contracting party.

Accession of Hungary to EU is seen by the members as an important link between East and West. The integration of Slovenia (an Adriatic port country) would further facilitate Hungary's access to the sea. Expansion of the EU and as a result, prevailing dynamics of trading opportunities between the eastern and the western borders of Europe have considerably reduced Hungary's disadvantages of being a LLC. For example, Hungary takes full advantage of this situation and its position by emphasising on service sector development and exploiting the advantages of its role as a transit country. Moreover, recently, Hungary managed to lure away high-tech investment from Ireland, so far considered as one of the best manufacturing sites in Europe. This is all the more interesting, as one of the reasons for this move was also Ireland's geographical situation that requires transhipment of finished goods overseas to the EU. Hungary's landlocked location, coupled with sharing of borders with seven countries and connections to the emerging economies eastwards was exactly what attracted investors.

North East : Isolated, Landlocked and Underdeveloped

In addition to the geographical distance separating the North East region from mainland India, the region is even farther away from the conscience of Indian leaders. Though hemmed in by a long chain of international borders, the region is linked to the Indian mainland by a slender 21 km. wide corridor through North Bengal. This not only make the North East region a strictly landlocked hinterland region of India but also render it very much isolated from the international community, thanks to the 'closed door' policy of India towards its eastern neighbours and the infamous 'Protected Area Permit (PAP)' system imposed in the region. Under the system, no foreigner is allowed to stay in the North-Eastern States beyond ten days. Moreover the foreign national is required to seek permission from the Ministry of Home Affairs, New Delhi to visit the region. Such policy has negated any prospect of tourism industry in the region even as the policy makers in New Delhi have repeatedly acknowledged its huge potential. Along with the policy of discouraging foreign tourists in the name of 'security' virtual absence of tourism related infrastructure like roads, hospitality based institutions such as hotel, restaurants, recreation centres, theme parks, gardens, holiday resorts and high-tech banking make tourism a non-starter.

Though the Indian policy makers have realised the necessity to integrate the country with the economies of South and South East Asia very lately and have taken up a modest initiative towards this end under its Look East Policy, the North-Eastern region has not been integrated into the mainstream economy of the country, in the strict sense of the term. This is well manifested in the limited connectivity the country has with its North-Eastern States. In its more than 50 years of independence, the Government of India is unable to provide rail connectivity to the North East region fully. So far, the only viable rail connection with other cities and commercial centres of India is only up to Guwahati. The roadways in the region are substandard and unfit for round the year service. Citing transport bottlenecks in the region, the Government of India is unwilling to locate any worthwhile industrial base or manufacturing unit in the North-Eastern States. Over the years, the stubborn reluctance of the Government of India to invest in the North East resulted in severe economic disparity in comparison to other parts of the country. This is reflected clearly in the meagre per capita consumption of 102.7 KWH electricity in the region as compared to all India average of 335.42 KWH. The North-Eastern States share just 2.4 per cent of the total surfaced roads of India [NEDFi (North-Eastern Development Finance Corporation Limited) Guwahati, 1998–99]. The seven North-Eastern States share just a meagre little 2.54 per cent of the total net gross domestic product of India (Government of India, 2003).

While the rest of India has direct or indirect access to the sea through standardised roadways and railways, the North East is virtually sandwiched and landlocked by mainland India, Bhutan, China, Myanmar and Bangladesh. Though cross border trade in the North-Eastern region with the neighbouring countries was flourishing during the pre-colonial period, the region became isolated and land locked in the post-independent period following closure of several ancient trade routes. The ancient land route from Peshawar to Parvatipur (now in Bangladesh) which extended till Assam was cut off after the partition of India and creation of Bangladesh later. Moreover, the ancient route linking Brahmaputra valley to the South East Asia and Yunnan province of China was never restored after the route was closed by the British rulers to safeguard the interest of steam navigation companies owned by British subjects and operating in the rivers of Burma. Even though the land routes between Myanmar and Manipur exist till today, these routes are no more than foot corridors except for the Moreh-Tamu line. The Indo-Myanmar Trade Agreement was signed as late as April 21, 1994 and border trade through Land Customs Station, Moreh was officially opened only on April 12, 1995 under which only 22 items were sanctioned for free trade between Myanmar and India. The Indo-Myanmar border trade through Moreh has almost come to grinding halt just after three years. Unsuitable policies, detrimental restrictions, absence of necessary infrastructure and lack of uniform trade mechanism between the two countries are some of the factors responsible for the sudden decrease in the volume of border trade through Moreh. In order to boost trade between the North East region and the neighbouring countries, government should provide necessary infrastructural facility at different trade routes, establish rail linkages, ensure free flow of road traffic, improve telecommunication links and most importantly free border trade in local produces should be promoted (Bhattacharyya, 2000).

Yunan : A Case of Developing Hinterland

Located in China's south west, Yunan has so far been an impoverished landlocked region with an inhospitable terrain. Peopled by 26 different minority groups including Tibetans and the Dai who are of Thai ethnicity, the province has intermittently been the focus of secessionist movements. In many ways, Yunan was similar to India's North East, underdeveloped, geographically isolated, and home to diverse tribal groups and far from the political capital. Yet, following a concerted effort on the part of the Chinese government to develop infrastructure, lure tourist and open up the province, the average per capita income has doubled over the last decade to just under US$ 1,000. Tourism is flourishing and contributed almost 20 per cent of the province's weighty US$ 45 billion GDP last year. Some 50 million tourists, including 3 million from overseas, visited the region in 2005. Facilitating their travel in the province is a slew of six-lane highways and railroads. Yunan also boasts 11 airports including an international one in its capital Kunming.

In the words of Jin Cheng, Director General of the Province's International Regional Corporation office, the Yunan provincial government has a very clear development strategy. It wants to make full use of the region's cultural and ethnic diversity to develop tourism. In short, Yunan is projected as China's gateway to South Asia and South East Asia. Since Yunan lacks a port, the province's focus has been on opening and developing its border trading routes, gradually creating a regional economy that transcends international borders. The Yunan Foreign Affairs Office is thus currently working on six major international projects including the Greater Mekong Sub-Region (GMSR), which comprises Cambodia, Vietnam, Laos, Thailand, Myanmar, and Yunan in China as well as the BCIM or Bangladesh, China, India and Myanmar Initiative. Yunan's budget for developing cross-border infrastructure was US$ 2.5 billion in the last five-year plan period (2001–05). The creation of excellent road, rail and air links among the member-countries of the GMSR have enabled them to be marketed together as a part of a single tourist package in Europe. Indeed, the Mekong area has, as a result, become one of the world's fastest growing tourist destination (Aiyar, 2006).

Yunan's total GDP is Yuan 208 billion (approximately) and its per capita income at current price is US$ 650 as compared to the per capita income of US$ 300 for the North-Eastern States of India during the same period. Aggressive economic reform has translated South West China into a significant economic growth zone in the last decade and as a result, the Yunan province has been growing at nine per cent since 1981, far exceeding the target growth rate of seven per cent. Tibet, another landlocked province of China, is growing at nine per cent per annum since 1991, and has also crossed the national GDP growth rate. One major source of revenue for Yunan and Tibet is tourism industry. In fact, the flourishing industry in Yunan has propelled rapid development of other tertiary sectors of industry, such as catering, transportation and trade. Since the implementation of the national policy of reform, the various nationalities in Yunan have strived to develop a flourishing economy by creating some pillar industries, and to participate in economic cooperation with investors from home and abroad, greatly strengthening the comprehensive economic power of Yunan, and thereby improving the infrastructure for developing foreign trade. Within an area of 200 kzm. around Kunming, the main lines of communication have been modernised, and the preliminary construction of road network leading to borders and other provinces has also been completed. The completion of the Yunan-Vietnam Railway have increased the length of rail lines to more than 2,000 km, forming a railway network linking neighbouring and coastal provinces. The civil aviation sector of Yunan has also made remarkable progress by building and extending nine international airports. With its abundant natural resources, strong industrial foundation, conducive infrastructure and good investment climate, Yunan has been drawing international investments from Taiwan, Hong Kong, USA, Japan, Thailand, Singapore, Germany, Britain, South Korea, etc.

Despite the similarities pointed out earlier with India's North East in terms of geography and population, it is apparent that the parallels between the two regions end here. The seven States that comprise North East India continue to languish in poverty. Per capita incomes are well below the national average. Traditional trade routes crucial to the local economies were rendered defunct as national borders have been closed because of India's nagging suspicion of its eastern neighbours. The whole area faces a debilitating paradox. Cross-border trade has been virtually outlawed due to security concerns even as trade with the rest of India through the narrow Siliguri corridor has failed to take off.

On the contrary, the North East region of India presents a dismal picture as far as development initiatives are concerned. In fact, the North East region is very poor in all aspects of infrastructure as compared to the all India average. Even though the North East is the only corridor between South East Asia and India, the region remains landlocked and isolated geographically and economically due to the Government of India's wariness towards its eastern neighbours. In stark contrast to the economic backwardness of the region, North-eastern India is bountifully endowed with rich bio-diversity, hydro-potential, oil and gas, coal, limestone and forest wealth. It is ideally situated for a whole range of plantation crops, spices, fruits, vegetables, flowers and herbs, much of which could be processed and exported to neighbouring countries. There is also a huge potential for the development of a prospective tourism sector. In reality, the North East region provides the crucial link to South East and East Asia (Verghese, 1997).

Though the notion of industrial backwardness is associated with the region, the fact stands that tea industry started as early as 1886 and is still thriving, and the oil refinery started in 1897 is the oldest in Asia. With 38 per cent of the country's hydropower potential located in the region, the climate for investment is most favourable owing to an almost negligible pollution and a high percentage of English speaking people. Some of the potential areas with immense potential to propel the North East region on to the path of economic prosperity are identified as agro and food processing, horticulture and forestry, energy, coal, natural gas and oil and tourism.

In fact, it has been rightly concluded that the region enjoys several advantages over other parts of India such as :

(i) An emerging market of 400 million people including neighbouring countries like Bangladesh, Bhutan, China, Myanmar and Nepal;
(ii) The potential of developing into India's powerhouse;
(iii) Unlimited tourism opportunities—rare flora and fauna, natural scenic beauty, unique performing arts, varied cuisine and textiles; and
(iv) Locational advantages, proximity to SAARC and ASEAN regions, making it attractive to foreign investment.

North East India's locational advantage and rich natural resources provide ample scope for its transformation into a base for cooperation with neighbouring countries like Bangladesh, Bhutan, China, Myanmar and Nepal. And through Myanmar, sub-regional cooperation can be extended to Thailand, Laos, Cambodia and Vietnam.

While the South West China including the Yunan province which has hitherto been very much similar to North East India in terms of geographical isolation, landlockedness and underdevelopment, has surged ahead with robust growth rate, the North East region, on the other hand, is languishing under poverty and underdevelopment. The point of contrast here is that the Chinese Government manipulated the location of Yunan and South West China to its advantage through multi-layer development strategies which include regional integration, infrastructure development, promotion of tourism and establishment of local resource based industries. In fact South West China and Yunnan have successfully negated the impediments of being a landlocked region. Even though the North East region is equally endowed with plenty of resources, it still remains a landlocked, underdeveloped hinterland region due to virtual absence of infrastructure, absence of conducive legal atmosphere and New Delhi's conspicuous lack of initiative to invest and build productive base in the region. It is appalling to see India's failure to use the region's highly advantageous position bordering Myanmar, China, Bangladesh, Nepal and Bhutan, which can naturally serve as gateway to South East Asia.

Conclusion

The basic argument of this essay is to drive home the point that landlockedness and underdevelopment are not inseparable. The obstacles faced by landlocked countries on the path to economic development are neither insurmountable nor underdevelopment is the destiny for landlocked countries or hinterland regions. Even as it has been accepted that LLCs and hinterland regions are challenged by certain constraints and disadvantages, the modern economic dynamics based on regional integration has raised the necessity to question the rationality of underdevelopment and landlockedness.

Recent studies of some developed and developing LLCs and hinterland regions have identified uniform transport policy encompassing transit states, financing investment in transport infrastructure, regional integration, harmonising transport documents and procedures and information technology as the basic tools. It needs more proactive role on the part of the State as is rightly stressed in the 'new trade policy'. For a landlocked country or a hinterland region, it is matter of great paramountcy to arrange viable alternative trade routes instead of depending on a single transit state as political uncertainty or social unrest in the transit state is most likely to produce adverse effects on the economy and commerce of the LLC.

Besides reducing in transport cost and time, improvised transport infrastructure has the potential to considerably allay the uncertainty factor faced by investors in relation to the costs of transportation through the transit country. Investment in transport infrastructure in LLCs may prove futile if the complementary infrastructure of the transit state is not adequate. Such problems arising out of non-complementary infrastructure and non-uniform transport policies and procedures can be effectively redressed through regional integration and harmonising transport procedures and customs. Establishment of connected market systems or common markets, reduction in manufacturing and transportation costs, effective coordination of multinational infrastructure projects etc. are some of the advantages of regional integration, most notably observed in the case of EU and ASEAN, thereby boosting the economy of LLCs in tandem with coastal countries. Another thrust area which is of extra significance to LLCs is research and development of information technology. Realising the inevitability of regional integration for economic development of a country, especially an LLC, the Yunan Foreign Affairs Office is working on six international regional integration projects including the Greater Mekong Sub-Region (GMSR), BCIM, the Kunming Initiative, etc. Moreover, the Chinese Government is exploiting the province's resources besides promoting tourism and encouraging investment both from home as well as abroad.

Though equally endowed with bountiful natural resources, the North East India on the other side of the border is still plagued by underdevelopment and remoteness. Despite the modest and late attempt of the Government of India towards regional integration through its Look East Policy, it remains highly doubtful whether the LEP would benefit the North-Eastern States as no serious effort is visible to invest or encourage investment for exploitation of the resources available in the region, nor there is any worthwhile infrastructure development. Not only that, the North East region has been forfeited of a huge amount of revenue from the tourism sector due to the infamous PAP (Protected Area Permit) system imposed in the region.

In short, the underdevelopment of the region is neither insurmountable nor its own making. The remoteness of the region is a result of insidious policy followed by the Government of India over the years in the region, and its wariness towards its eastern neighbours has only consolidated remoteness of the region. On the contrary, the region by virtue of its location in the middle of the trade route connecting Indian subcontinent and South East and East Asia, the North East India has immense potential to become the commercial hub of the entire Asiatic continent.

REFERENCES

Commission of the European Communities. 2001. *Energy and Transport in Figures: Goods Transport, Other European Countries, Performance by Mode in 1998.* Brussels: Commission of the European Communities.

Gallup, J.L.; J.D. Sacks and A.D. Mellinger. 1998. *Geography and Economic Development.* NBER Working Paper No. 6844.

Glassner, M I. 1992. *Bibliography on Landlocked States.* Third edition. Dordecht, Netherlands: Martinus Nijhoff.

Government of India. 2003. *Economic Survey 2002–03.* New Delhi: Directorate of Economics and Statistics.

Healey and Baker. 2001. *European Distribution Report.*

Jakob L., Annovazzi. 2003. *Landlocked Countries: Opportunities, Challenges, Recommendations.*

Krugman, P. 1987. Is free trade passe? *Economic Perspectives* 1, 2: 131–44.

Limao Nuno, Venebles Anthony J. 2001. *Infrastructure, Geographical Disadvantages and Transport Costs.* Washington DC: Columbia University and World Bank.

Mackellar, L., J. Worz and A. Worgotter. 2000. Economic development problems of landlocked countries. *Transition Economics Series* 14. January. Vienna: Institute for Advanced Studies.

Smith, Adam. 1976. *An Inquiry into the Nature and the Causes of the Wealth of Nations.* Chicago: Chicago University Press.

UNCTAD. 2001. Specific actions related to the particular needs and problems of landlocked developing countries. Note by the Secretary General (A/ 56/427) in General Assembly, October 10, in New York, United States of America.

World Bank. 1998. *World Development Indicators.* 1998. CD-ROM. Washington DC.

World Bank Operations Evaluation Division. 1994. *World Bank Lending for African Transport Corridors: An OED review.* Washington DC.

2

Geo-Spatial Reality and Development Paradigm : A Case of the North East Region

Konthoujam Indrakumar

Resounding cries for 'development' in the North East region is gaining ground with increasing consciousness amongst the masses. The seriousness with which peoples in the region look at "development" issues suggests the intense urge for meaningful, holistic and inclusive development in the region. Urge for development is reflected in the growing debates and corpus of literature on economic development in the region. Intensive focus on the region's economic potentials and *modus operandi* of the Centre's development undertakings bring to light many crucial issues. Even though the agenda emerging at the bottom level of the movements articulated *vis-à-vis* growing aspirations cannot be suppressed and marginalised without generating conflicts, the issue of development has indeed opened whole new perspectives of understanding the political economy of the region. It is a historically ascertained truth that in every socio-political movement political goals are always associated with economic realities of the struggling communities or nations. Priority objective of the movement is conditioned by the material reality and circumstances.

Coming to the development debates in the North East, development in the region remain as illusive as peace. At the same time State initiatives for development are highly charged with different security-centric perspectives and policy formulation and implementation remain highly centralized in command and structure.

The incompatible propositions advanced by the 'civil society bodies' on the one hand and the State represented by politicians and bureaucrats on the other hand continues to exude conflicting positions regarding development in the region. Current debates are shrouded with paradoxical and often-contradictory claims. For instance State claims that the region remains backward on account of the disturbing law and order situation created by insurgency. The statist proposition of ascribing the region's backwardness to insurgency is based on the premises that prevalence of armed opposition groups lead to siphoning off huge amount of fund meant for development works and 'disturbances' acting as stumbling blocks to successful implementation of development programmes. Ironically those in power also acknowledge that economic backwardness or lack of development has become the major propelling factor behind the genesis of insurgency thereby assuming that without peace development in the region is virtually impossible. Such argument of development in the region merely creates a situation of 'vicious circle' and pave the way to gain political legitimacy for further militarization in the region.

However, unlike the statist proposition, civil society bodies and many intellectuals' argument on the region's development discourse take a reverse stand. Besides, it also helps in different premises. Their argument is that peace is not merely cessation of fire or the absence of conflict; it has to be based on fulfilling aspirations by addressing the genuine grievances of the masses in the region. Justice in their view forms the main plank of peace in the region. In order to bring peace, an integrated development strategy *sans* militarization is required. Hence, the slogan 'development and peace' rather than 'peace and development' has been articulated rigorously. One of the logical outcomes of the factors cited by the State for underdevelopment of North-Eastern region is the massive military build-up for counter insurgency operation. In fact more than 50 years of fighting insurgency with India's full military might supported by one of the most draconian legislations in human history known as the Armed Forces Special Powers Act (1958) has failed to yield intended results. In fact it has only exacerbated and aggravated the problem. The Act has only neither been able to contain proliferation and growth of insurgency in the region nor restore peace and tranquillity in the North East to create the 'conducive atmosphere' which the State claims as essential for successful implementation of development programmes.

Therefore, a new premise with an alternative framework is crucial to address the underdeveloped condition of the region that can lead to a new kind of strategy for developing North East. Instead of looking at the problem in the region from the "law and order" perspective, there is need for going beyond it and set new perspectives to perceive the problems in its historical perspectives. Observers and analysts of the region point out the futility of treating the issue of insurgency in the North East simply as "law and order" problem. It is not the solution, economic and political causes of insurgency have to be recognized and dealt with properly. It is not within the purview of this essay to delve into the North East and growth of insurgency or armed opposition movements in the region.

The essay aims to address the issue of underdevelopment of region by trying to comprehend the reasons and provide on alternative frame of reference. It is a well-known fact that the region in spite of possessing huge potentials for development remains still backward. Despite huge potentials to generate economic development, which can subsequently contribute to the greater wellbeing of the population in the region it is still steeped in immense poverty. In short, the region is replete with abundant forest resources in terms of its varied natural flora and fauna, minerals, limestone, coal, uranium, oil, natural gas, etc. With more than 300 varieties of medicinal plants and herbs and 400 species of wild orchids, the region is unique in terms of being a bio-diversity hot spot. Besides, the region possesses :

1. Vast water resources including hydel power potential with more than 50,000 MW identified capacity;
2. Tourism potential;
3. Proximity to South East Asia, one of the world's fastest-growing economies;

4. Rich heritage of handloom and handicrafts; and
5. A democratic traditional system of local self-government with community spirit permeating the entire social system (Khaund, 2006, 14).

In spite of above potentials, which otherwise can propel the region into a developed one, it remains a 'dependent periphery'. Despite pumping of huge amounts of central's funds development in the region still remains a distant dream. For instance between 1992 and 2002 alone, central fund totalling close to Rs. 38,000 crore under Five Year Plans, Border Area Development Plan (BADP), North-Eastern Council (NEC) grants, Non Lapsable Central Pool of Resource (NLCPR), etc. released to the States in the North East, have failed to generate development in the economy of the region. There seems to be many unanswered questions hidden beneath the development debate in the region. Whether underdevelopment in the region has to be attributed to its geographical feature or to the location and geography purportedly ascribed as crucial factor in making the region remote and isolated or is it historical and political factors that need to be critically analysed. It is being increasingly argued that history and politics rather than location and geography subsequently contribute to the relative underdevelopment of the region. Development discourse, therefore, must incorporate the geo-political and historical factors including the attitude of the central leaders towards the region along with the development policy pursued so far. If one accepts development as a cumulative and concomitant outcome of long years of consistent hard work and planning along with deep understanding of the socio-political contexts of a region, it is here that the centre has specifically failed *vis-à-vis* the region.

Location and Development : A Fresh Look

Pervasive tendencies to correlate the economic status of a region with its physical geography persist in classical theories of development. It has been the crux of rigorous debates both at the national and international levels. A cursory glance at the map shows that territorial space referred to as the North East is demonstrably landlocked without any outlet to the sea for trade and commerce. This cartographic fact usually sends out the message that the region's economic underdevelopment is mainly because of the limitations arising out of this undeniable geographical fact. Geographical reality of the North East and its economic status sum to be rigorously pursued in order to legitimise its underdeveloped condition. As asserted by analysts that this undeniable fact and perception of the disabilities and limitations arising out of this geographical reality have affected economic growth of the region, creating some unique security problems deeply entrenched in the region, is passionately articulated both at the popular and professional levels. Discussion of the problems of development faced by landlocked states or even territories of a country which itself is not landlocked, as in the present context assumes that there is a near axiomatic correlation between its landlocked position and its economic underdevelopment. This correlation appears so glaring that one tends not to think outside this correlation, or see the causes for the undeniable economic and social backwardness of such landlocked territories in factor other than physical geography (Prabhakara, 2004).

To be fair to the proponents of such correlation between locational features and economic and social backwardness, the nature of geographical features can hinder development efforts due to difficult terrain. It also accrues higher cost in terms of infrastructure development and higher transportation cost. However, such impediments are of curative nature as experienced elsewhere in India and abroad, given the firm political determination and will to develop along with application of proper technology. Emerging realities from different landlocked states in the world are not always uniform in terms of progress and development in economic and social aspects. A survey of the landlocked states of Europe, their prosperity and development, falsify the demonstrable relationship between locational features and development status of a state or a region. Being landlocked does not deny the possibility of developing a region or a state; citing land locked character as 'sole determinant' for underdevelopment of North East India carry the danger of overlooking various other factors for underdevelopment in the region. Need and aspirations of the region's peoples can be fulfilled only through proper planning that addresses the issues of productive industry based on utilization of the region's resources to the advantages and benefits of its peoples.

To further elaborate the point a study of the experiences of landlocked states at the global level 'unlock' this correlational factor between geographical location and underdevelopment as an artificial construct rather than an incorrigible natural impediment which always serve as stumbling block to development. It can be easily discerned that the European landlocked states like Switzerland, Austria, Luxemburg and the Czech Republic are at par with other European states in terms of human development index. They are among the most developed and prosperous countries in the world. An interesting fact is that seven landlocked states of Asia, part of the erstwhile Soviet Union, are well placed above other landlocked Asian states like Afghanistan, Bhutan, Laos and Nepal which come at the bottom of the Human Development Index (HDI) listing (UNDP, 2004). This clearly indicates the unequal progress in terms of development among the global landlocked states. The reason for these variations has to be located in the greater socio-economic environment of the states concerned and particular politico-economic policies pursued by these states at different stages of their historical evolution.

To elabsorate the point a little lengthier one can cite the examples of Yunan, a landlocked province of China and Macedonia, a small landlocked state in Europe. Yunan has a striking resemblance with India's underdeveloped North East region in various aspects. Yunan, located in the south west of China, has for long been an impoverished and underdeveloped region with an inhospitable terrain. 26 different minority groups populate the province. Besides, the province had been a centre for armed movements. In various ways Yunan—historically underdeveloped, geographically isolated area, home to diverse ethnic groups and far from the political capital of Beijing—was a replica of India's North-Eastern region. However, following a concerted effort to develop infrastructure, lure in tourists and open up the province to bordering countries, the average per capita income of Yunan has doubled over the last decade to just under US$ 1,000. Flourishing tourism contributed almost 20 percent of the province's weighty US$ 45 billion gross domestic product (GDP) in the year 2005. Some 50 million tourists, including three million from overseas, visited the region in 2005. The consistent policy of Yunan government based on a very clear development strategy having multipronged agenda like infrastructure development; cross border diplomacy and strategic marketing are important factors behind the economic boom of historically underdeveloped Yunan Province of China. Yunan Government has been able to convert instable condition entrenched in multi-ethnic, diverse cultural traditions into its advantage by making the polyethnic and cultural diversities a hot cake for tourism. The newly constructed six-lane highways and railroads facilitated to a great extent in developing the province by expanding trade and commerce. At the top of all these achievements Yunan Foreign Affairs office is currently initiating development strategy in the Mekong Sub-Region (GMSR) comprising Cambodia, Vietnam, Laos, Thailand, Myanmar and Yunan and the BCIM or Bangladesh, China, India and Myanmar, initiative, Yunan's budget for developing cross-border infrastructure was RMB 20 (US$ 2.5 billion) in the last five year plan period (2001-2005) (Aiyar, 2006, 63–4).

It is because of the development miracle of Yunan that professionals and intellectuals in the region began to anticipate the possibility of mapping Yunan's model for the North East because of the striking resemblances between the two regions. Development taking place in Yunan was not borne out suddenly but a result of consistent development strategy with strong intervention from Beijing. Beijing granted much needed political autonomy to the provincial government of Yunan to reflect on crucial issues, formulate and execute policies and strategies for its own development. Offices such as Yunan Foreign Affairs office, Yunan Development Research Centre, Yunan's International Regional Cooperation Office, etc. are important institutions that are contributing remarkably in the growth and development of the province, which are conspicuous by their absence in the case of the North East. These institutions play crucial roles in formulating development strategy negotiating with its neighbours for trade and commerce including infrastructural upgradation. Freedom to pursue an independent economic policy *vis-à-vis* its neighbours and formulation of development strategy with proper planning perspectives are crucial in transforming what are hitherto challenges into opportunities thereby enabling its government to translate potentialities into reality. In this respect the North East region remains far behind. With the exception of geographical and demographic similarities with Yunan the North East region is engulfed in the stark realities of a 'dependent periphery', a 'distant frontier'--both in terms of geographical and psychological perception—and a 'military buffer zone' in the geo-political environment of India and China.

A Case of Macedonia : Story of Survivors

A brief reflection on the geography, polity and economy of Macedonia reveal the political will and human ingenuity for crossing artificial and natural barriers to develop and survive by overcoming the challenges. Macedonia, a landlocked nation in the mountainous heart of the Balkan Peninsula, was one of the six republics of erstwhile Yugoslavia federation. Macedonia, a small country having 25,333 sq. km, a little over the size of Manipur State in India's North East is populated by various ethnic groups, such as Macedonians (64.18 per cent), Albanians (21.17 per cent), Turks (3.85 per cent), Roma (2.66 per cent), Serbs (1.78 per cent), Vlachs (0.48 per cent) and others (1.04 per cent). Ever since the country became politically independent from the erstwhile Socialist Republic of Yugoslavia in 1991, through a referendum on 8th September, 1991 endorsing independence, it has undergone the pangs of sudden break-up from its parent federation including political instability caused by the assertion of ethnic Albanians for autonomy, border disputes with its neighbours, controversy over the issue of nomenclature with Greece leading to trade embargo and other forms of disturbances.

However the resilient Republic of Macedonia survived through these upheavals. The plight of Macedonians and its uncertainty as a sovereign political entity is reflected in this complex and fractious bottleneck on the edge of Europe, each nation dreaming that it may someday regain the borders of its empire at its apex. These borders have often lapped across one another (Vesilind, 1996, 126). In fact each of the neighbours of Macedonia has a certain grudge against the country. Greeks are obsessed with the name Macedonia. They consider it as a distortion of Greek history and symbol. As a result they closed the major trade artery of Macedonia from its port of Thessaloniki to Belgrade and beyond. On the east the Bulgarians tend to patronize Macedonians, claiming they are merely western Bulgarians with nationalist delusions. To the west tens of thousands of Albanians who have lived in Western Macedonia for generations hope to tear off a chunk of the new nation as an autonomous republic for themselves. To the north militant Serbs consider Macedonians misguided country cousins who belong to a Greater Serbia (Versilind, 1996, 126).

Economically Macedonia suffered a series of set back in the early 1990s of the twentieth century due to the emergence of intra-federation war between Serbia-Croatia, Serbia-Bosnia as a result of which Republic of Macedonia became independent. One of the economic consequences of war was the imposition of blockade against Serbia by the United Nations. Economic blockade against the Serbia affected the economy of Macedonia because some 62 per cent of its trade was with Serbia. Economically it was described as a poor country where light industries, tourism and agriculture were the mainstay of its economy. Besides, trade embargo by Greece brought severe economic hardships to the country. Along with it Macedonia's enforcement of UN sanctions against Serbia was said to have cost another US$1.9 billion per year, according to Macedonian officials (Perry, 1996, 116). As a result, unemployment was high; it was 2,22,804 in August 1995 out of the state's multiethnic population of just over 2 million (1994 census). Some of Macedonia's 400,000 employed workers did not receive regular pay. However much water has flown since 1991 in Vardar River; with consistent efforts Macedonia is on the path of economic recovery and generation. The country has been able to maintain macro economic stability with low inflation. Today the Republic of Macedonia is considered a country with an intermediary-developed industry and continuing growth of the industrial production.

With restoration of trade relations with Serbia and Greece and opening up of major trading routes, the country's future seems stable. There is every possibility of development taking place at a faster pace. The Republic of Macedonia, even though small in its territorial aspect, has been able to survive because of its productive base like food processing, beverages, textile, chemicals, iron, steel, cement, energy, pharmaceuticals which form the main industrial bases of its economy. Its exports commodities consist of food items, beverage, tobacco, textile, miscellaneous manufactures, iron and steel. The country's main exports partners are Serbia and Montenegro (22.5 per cent), Germany (17.8 per cent) Greece (15.3 per cent), Italy (8.3 per cent) in 2005. Agriculture contributes 9 per cent, industry 29 per cent and service sector 62 per cent to country's GDP in 2006 (http//www.cia.gov/cia/publications/factbook/geos/mk.html).

Correlation between location and development cannot be generalized and landlockedness as a sole factor of economic backwardness is a contested and controversial one. As argued above development experiences in landlocked states are not similar in many aspects. Therefore, the condition of underdevelopment must be attributed beyond the geographical feature of a country. It should also be attributed to the factors like socio-economic environment, perception of those at the helm of the affairs, policy and programmes they have undertaken.

Development Experience in North East : Need for Rethinking

The above analysis of the case of Yunan Province of China and the Republic of Macedonia is intended to focus on the contention that farsighted development strategy can overcome hindrances posed by geographical features of a country or a region. Formation of such development strategies involves the contextual understanding of a region's socio-political landscapes and exploration of its relatively stronger conditions. Yunan province of China and Republic of Macedonia depict striking similarities in terms of geography, ethnic profile and political instability with the North East. However the similarity ends only with geography. The North East region remains relatively an underdeveloped, a historically deprived in India and dependent periphery amidst huge resources and potentials. With poor industrial base the region is solely dependent on the rest of the country for variety of required commodities. Underdeveloped condition of the region can be discerned from the fact that the percentage of industrial production to GDP is 2.16 in the region against the all India average of 7.7 percent. Per capita power consumption in the region is 97 KWH against the all India average of 355 KWH. It depicts the gloomy picture of the region's economy. However lack of development in the region cannot be solely attributed to its topographical feature because the States having sea outlet in India such as Orissa, Telangana region of Andhra Pradesh are relatively backward than the landlocked States like Punjab, Haryana, Himachal Pradesh, and Rajasthan. More interestingly more developed States of India like Punjab and Haryana are doubly landlocked (Prabhakara, 2004, 2911–916).

In the case of North East the dynamics of history, politics and perception of New Delhi towards the region influenced by certain geo-political development in the subcontinent and beyond have played major roles in transforming the region into a less developed one. Economic implications for having 98 percent of North East territorial boundary with neighbouring countries is yet to be transformed into a real possibility. It is entirely connected with the rest of India through remaining two percent of its territory for various purposes. The geo-political development in the subcontinent during the period following British departure and subsequent partition of the British Indian Empire into two countries i.e. India and Pakistan and attempt of both India and Pakistan to incorporate many princely states in their fold by means of political persuasion and military intimidation continue to haunt the region even after more than five decades of India's political independence. If we look at the pre-British period history of the region, we can establish the fact that present day North East region was well connected with the outside world and there were trade relations among the kingdoms of the region and beyond. Existence of various trade routes is a testimony to the fact that flourishing trade between the kingdoms of the region and its neighbouring kingdoms existed in the past. D.G.E. Hall mentions about one such road linking lower Burma and India via the bank of the Irrawaddy, the bank of the Chindwin and the Manipur. There were at least three routes connecting Surma Valley with the Manipur Valley and two connecting the latter with the Kabaw Valley in the Chindwin river Valley in the present day Myanmar. The region produced many items, which were marketed outside the region. Silk cloths, ivory products, etc. from Kamarupa continued to be appreciated outside the region. Many items were imported from China, Tibet and Myanmar (Burma) through this region and exported outside India from this region (Kumar, 2000, 4–5). Again Pemberton in his report on *The Eastern Frontier of India* mentions that :

During the reign of Choorjeet Singh, a very brisk trade was carried on with the Burmese inhabitants of the frontier through the Kubo Valley, and Chinese merchants from Yunan used sometimes to visit Munepoor for purposes of traffic, bringing with them the same description of articles they take to Ava and carrying away, the wax, ivory, cloths, cotton and ponies of the country (2005, 33).

Commenting on the Pre-British period economic activities of the small princely state of Manipur Pemberton in his report also mentioned, 'with the exception of woollen clothing, the country itself furnishes every article that may be considered essential to the comfort and prosperity of its inhabitants' (Pemberton, 2005, 35). However, during the British colonial period the traditional trading pattern and economic activities of the princely states of the region was disturbed to the detrimental of economic health of the states in many ways. British colonization led to the economic exploitation of the region especially tea, oil, timber from Assam and other states of the region. The region remained by and large a supplier of raw materials and market for colonial power. Destruction of the region's productive base during British period also led to the intrusion of non-indigenous settlers who were basically traders from the Indian heartland. Despite British colonial exploitation, the region was able to maintain self-sufficient nature of its economy.

However, in the post-British period, development trajectory in the region is characterised by certain distinctive features, such as continuation of colonial legacy, large-scale migration influx monopolizing trade and commerce and constant militarization of the region. During the first three decades of independence little was done to develop industry, transport, banking or suitable means of communication in the region (Dutta, 1997, 46). The argument can be substantiated from the case of Assam. In 1950-51 on the threshold of the planning era, Assam had a per capita income higher than the national average by Rs. 50.4. In the course of three and a half decades of planning the State not only lost its initial edge but also continued its steady downward slide much below the national average. The disadvantages of being a landlocked region resulting from the partition of the subcontinent have remained intact and there seems to be no serious effort to reverse the situation (Indrakumar, 2006, 53).

Tragedy of post-British period is the large-scale migration influx from outside and consolidation of Marwari traders in the region thereby draining huge wealth from the region. Influx of labourers and traders from different parts of India and neighbouring countries has led to a massive demographic change and pressure on the land and resources of the region thereby affecting the socio-economic well-being of the region's indigenous population. The outsiders' contribution to the region as far as its economic well-being is concerned is negligible since they merely act as parasites in the body-economy of the region. If everyday experience and social reality, aside the academic discussions, has any meaningful indications then the fact of migrants (both from India and outside) controlling trade and commerce and cheap labour penetrating into the region's economy constitute a great source of insecurity and frustration. Failure to cope up with this problem on the part of the government is a great source of harm to the region's economy and social landscape.

Development scenario in the region clearly depicts the helpless and restless situation of the region's wholesome pathetic condition while its economic life is dominated by migrants and non-indigenous business communities, the political structure controlled by a centralised authority, its institutions being strangled by a group of elite class, with relatively poor power-base in society (Dutta, 1997, 44). To be more precise out of 620 tea gardens in the Brahmaputra Valley, only 158 small gardens were owned by the Assamese but the non-Assamese belonging to top seventy-five business groups of India owned 248 large gardens and factories with outside companies owning the rest of the large gardens (Misra, 1988 quoted in Dutta, 1997, 46).

Persistence of such patterns only erodes the already fragile economic conditions of the region. The situation in other States of the region is no better than Assam where most of the trade and commerce activities is controlled and dominated by the non-indigenous traders coming from different parts of India. Development policies prevailing in the region only subserve the interest of mainland traders. It can, therefore, be argued that the extractive nature of the economy practiced by the British did not undergo any major change in the first few decades of independence (Misra, 2006, 7).

Another dimension of development trajectory in the region is the ever-increasing dependency level of the region. It may not be an exaggeration to assert that more than five decades of India's planned development strategy has only led to enhance the dependency level of the region to the rest of India and has failed to generate much needed inertia for internal development in the region with the exception of a few facades of tall office buildings and concrete walls in selective urban centres. The Shukla Commission Report mentions that about Rs. 2,500 crore worth of consumables is imported from outside the region every year. Besides, the unaccounted unofficial trade is also rampant in the region. The report only indicated the undeniable fact of accentuating dependent status of the region even for daily essential commodities.

All the North-Eastern States depend on central grant in aid for their sustenance and development work. It is pertinent to mention that seven States of North East, along with Himachal, Sikkim, Jammu & Kashmir and Uttaranchal are termed as 'Special Category States'. These States receive the Planning Commissions' assistance in terms of 90 percent grant and 10 percent loan, contrary to the rest of the States' 70 percent loan and 30 percent grant (quoted in Agarwal, 2006, 151). In other words, the North-Eastern States are 90 percent dependent upon central assistance and most of the States are unable to repay ten percent share of the loans. Pathetic nature of the region's economic backwardness is reflected from the fact that most of the States in the region are fiscal deficit States right from 1987 to 2000 as mentioned in the Eleventh Finance Commission Report (2000, 182–183). Continuous pumping of huge financial packages neither generates employment nor alleviates poverty in the region. Unemployment level in the region is 12 percent against the all India average of 7.7 percent. The level of poverty in the region is higher than national average of 26.1 percent (Government of India, 2006).

Nowhere in India except in the North East Region and Jammu & Kashmir did the insecurity complex of the ruling clique in New Delhi play a tremendous role in shaping the perception towards the region and formulation of development policy. It can be discerned from the record fact that when there was demand for the setting up of refinery in Assam by the Assam Government, Nehru wrote to the then Chief Minister, B.R. Medhi, that for security reasons, such a refinery could not be set up in the State. Hence it was decided to pump out the crude oil from Assam to Barauni in Bihar by laying a 1500 km. pipeline involving astronomical costs and *depriving benefit that can be accrued from the refinery to its people* (emphasis added) (Misra, 2005, 6). The often quoted infamous Nehru comment on Assam in 1962 Sino-Indian war, and imposition of AFSPA 1958, a British colonial residue, are indicators of the fact that the insecurity complex dominates the New Delhi's mindset towards the region. Still New Delhi fails to come out of this mindset. Huge development packages associated with the constant military build-up indicate something, which is incomprehensible to the minds of masses and that does not augur well in fetching the confidence of the peoples in the region.

When one looks at the recent political historiography of the region one is tempted to say that New Delhi's attention to the region was always drawn when there is violent agitations. It is an undeniable fact that New Delhi began to view the region seriously only after the States in the region are directly challenged by the armed opposition groups of the region. Still the Centre's perception towards the region is deeply embroiled in security considerations. Therefore, New Delhi needs to change this attitude and there is need for dissecting development strategy from military and geo-strategic considerations so that development occupies a neutral and autonomous notion of its own.

Conclusion

Introspective analysis of the development effort in the region reveals increasing dependency without generating an inbuilt strength for regional development planning. Superimposition and centralised system of planning only serve the interest of the few urban class comprising of politicians, bureaucrats and contractors in the region sacrificing the larger interest of the peoples. Central development efforts have failed to bring rural-urban parity in the region. Development in its holistic sense can only be brought in the region by creating necessary productive bases based on regional resources. Efficient utilization of the region's resources to the advantages and benefits of its indigenous populace should be an immediate strategy for development in the region. Growth of tertiary sector without a significant increase in the share of primary and secondary sectors indicates that commodity producing economic activities are still unable to render their due share in the economic development of the North-Eastern States. Without significant growth of primary and secondary sectors the momentum of growth in the economy cannot be sustained longer. The dimension of migration influx and their impact on the economy of the region need to be assessed properly and corrective measures should be initiated. Underdevelopment of the region has to be located within the broader socio-economic environment and factors inherited in the historical evolution of the region. The difficult terrain and physical handicap arising out of the topographical features are not the root cause of underdevelopment and development discourse in the region should go beyond geographical factors and insurgency.

REFERENCES

Agarwal, A.K. 2006. Flow of central funds and its impact on North-East Economy. In *Challenges of Development in North-East India*, eds. David R. Syiemlieh, 146–75. New Delhi: Regency Publications.

Aiyar, Pallavi, 2006. Yunan model. *Frontline* 23. October 21–November 3: 63–5.

Dutta, Anup. 1997. Peripheral States, underdevelopment and identity crisis: Conflict and integration in Northeast. In *Politics of Identity and Nation Building in North East India*, eds. Girin Phukon and N.L. Dutta, 45–58. New Delhi: South Asian Publication.

Government of India. 2006. *Annual Report* 2005-2006. New Delhi: DONER.

Khaund, T.P. 2006. North East at crossroad. *Yojana* 50. December: 13–5.

Kumar, B.B. 2000. The border trade in North-East India: The historical perspective. In *Border Trade North-East India and Neighbouring Countries*, ed. Das, Gurudas and R.K. Purkayastha, 3–11. New Delhi: Akansha Publishing House.

Misra, Udyon. 2006. Ethnicity, territoriality and autonomy in India's North-East: From fragmentary politics to an inclusive social space. In *Challenges of Development in North-East India*, eds. David R. Syiemlieh, 1–11. New Delhi: Regency Publications.

Pemberton, R.B. 2005. *The Eastern Frontier of India*. New Delhi: Mittal Publications.

Perry, M. Duncan. 1996. Macedonia Balkan miracle or Balkan disaster. *Current History* 95, 599. March: 113–17.

Prabhakara, M.S. 2004. Is North East India landlocked? *Economic and Political Weekly* XXXIV, 9. October: 2911–916.

UNDP, 2004. *Human Development Report*. New York: Oxford University Press.

Versilind, J. Pritt. 1996. Macedonia: Caught in the middle. *National Geographic* 189, 3. March: 120–39.

3

Relocating the North East in Global Economic Space : Dynamics of a Periphery

Sanjeev Thingnam

This essay is a critical engagement with the North East in terms of its location and transformation as a space in the global economic mapping. It aims to study the developmental politics undertaken by the Indian Government in the light of ideological contours closely related to the idea of the region as a specific location and space for intervention. The politics of defining the region as part of an economic growth zone needs a critical inquiry in the light of recent policy perspectives making location and space relevant for defining developmental initiatives. Development involves many factors but location and space seem to have remained central in debates on the region. The essay attempts to study the evolution of developmental policies in the region in the backdrop of intractable political conflicts. Recent policy initiatives postulate possibilities of political integration of the North East region with mainstream Indian polity and economic integration with South East Asia. It is based on the belief that this will ease the region's problem both politically and economically. Such official views can be seen as a manifestation of the problem behind political integration faced by the successor Indian State after the British transferred power. There is widespread contention that the politics of defining the region has been from the same with the British Raj to the Indian Raj based in New Delhi. To talk about the continuity and changes at the policy recommendation level, both from the British and post-British framework, there is more of perceived continuity than change. It is crucial to understand the structural continuities and changes both at the politico-geographical level with its associated logic of intervention into that space to further developmental politics. Before going into the issue of policy formulation, it is important to know the various trajectories and attributes involved in the process of defining and conceptualizing the region.

Delineating the Political Geography of the Region

Conceptually a region is characterized by its location, area and limits describing the variable characters dissected deliberately from the extensive land mass. It is an artificial construct to uncover the particular physical and socio-cultural attributes of an area. Among the significant factors on which a region is identified, mention may be made of physiography, ethnicity, socio-political and economic systems, language, religion, degree of mobility, cultural-historical patterns and many more. Territoriality is an important attribute in defining a region. It is also a geographical concept indicating a variety of features. Population is another significant attribute of a region—a multitude of communities co-existing in the given territory over a period of time. Examining the idea of the North-Eastern region as a distinct entity, the consciousness of core (centre/mainstream) and frontier (periphery / margin) has remained rooted. Evolving relationship between the Indian core and the North Eastern peripheral region needs careful examination. The pattern of interaction between the two in future will determine the future course of history of the region (Sinha, 1994, 228). Exchange of human interaction has been there in this region since time immemorial. Economic, social, cultural and linguistic links between this region and South East Asian peoples have been a long-standing historical feature. General mainstream positions maintain that because of geo-political disadvantages, the region has achieved an integrated insularity. Its overall underdevelopment is an aspect of the larger scenario of unequal and uneven development in the various parts of the Union (Sinha, 1994). On the other hand, policy makers are considering the possibilities of transforming the region into a thriving integrated economic space linking two dynamic regions of Asia with a network of highways, railways, pipelines, and transmission lines crisscrossing the region. An apparently dramatic shift, it tries to envisage the North East not in terms of a locational periphery anymore.

North East as a Frontier Space and its Proposition

Notion of 'frontier' indicate two things—one is an area and the other a boundary line. However, the politics and purpose of defining this boundary line and its consumption in the 'mainstream' mindset are crucial issues that need to be delineated. A conscious attempt to impose a national and internal boundary in the post-British period perpetuates some of the intractable conflicts in the region (*Intercultural Resources and Forum for Indigenous Perspectives and Action*, 2006, 5). Subsequently, contestation arises in the process of restructuring the 'frontier' map that left many unresolved issues in post-British India. The politico-geographical spatial contour of the North-Eastern Frontier of India illuminates alternative frames of different spatial setting in different period of cultural continuum, time, and place context.

Colonial encounters with frontier inhabitants produced a rudimentary outline of political and cultural mapping. Patterns of reconnoitring and surveying the unknown terrain embedded in colonial projects like military expeditions and anthropological research led to redefining undefined territory of fluid frontier boundaries. Spatial dimension of the social and cultural domain of the frontier populace leaves a doubtful texture in the intellectual life of the 'National' Map. Desire for producing a new national political map after 'partition' is contested at large. For instance, maps of contemporary Indian social patterns extend to cover the adjoining and closely related cultural spaces of Pakistan, Bangladesh, Sri Lanka, and Nepal while covering the North East remains problematic. Similarly, in the case of the North-Eastern frontier, historical linkages with countries like Myanmar, Laos, Thailand and China seem to be undermined in the making of the Indian state. Conceptualization of the North Eastern frontier region as a politico-military space created many categories for exclusiveness where the spatial has remained a loose image for a perceived or desired racial and cultural divide. Apparently this kind of colonial project has a deep impact in post-British policy framework. Polity of the North Eastern frontier region was fluid in character and the relationship between tribes and kingdom was not of permanent subordination. Politically it was a kind of tributary relationship in terms of annual affairs or simply rather a *quid pro quo* relationship. The kingdoms followed a non-intrusion or non-interference policy. A larger paradigm shift came when the British paramount power endeavour to open up the tribal areas for direct communication to secure and stabilize the geo-politics of the region.

As a result the non-intrusive relationship of the local kingdoms and the surrounding tribes were affected. Apparently, the reason behind implementation of this policy could be the interest of the paramount authority to put the tribal areas permanently or semi-permanently under British suzerainty. In this context, two important reasons can be mentioned—firstly, the land route to Burma through Silchar was not favourable since Kohima *via* Angami territory proved. Thus the demarcation between tribal land and settled land became absolutely important for larger imperial interest. Secondly, delimitation of tribal and settled lands in absolute terms had to be worked out in such a way that it subserved colonial interest. Conspicuously, the tribal polity was compelled to enter into a permanently subordinate relationship with the emerging colonial administrative paraphernalia.

Discourse on Imperial Policy

Nineteenth century colonial policy fervently explored means and ways of exercising control over the Cachar Hills, north of the Water-pent, which had already been incorporated to Assam and placed under Nowgong district administration. To collect a house tax on the subordinate Nagas, and to station thannas in the hills, the Nowgong authorities proposed to make the hills a substantive district, under a separate administrative officer. However, it turned out that the whole return would amount to Rs. 3,000 annually, against an expenditure of over Rs. 16,000. Ultimately this scheme was not approved. Captain Jenkins, the Commissioner, proposed that the tract should be attached to Nowgong with an extra assistant official (Political Proceeding, 1839). He advised another official Mr. Grange to visit the Angami territory by the Samoogoodting route, to obtain the submission of those chiefs who might be willing to be subject to British Raj, and place a *thanna* of Shans on the Dhunsiri.

In this course of action, authority to subjugate and punish villages that resist the Paramount power was sanctioned. Captain Jenkins was insisted upon to fix the boundary line between Assam and Manipur and Assam and Cachar, which may be now clearly defined to be the Water-pent of the great ridge of mountains. He suggested that Manipur should be compelled to cooperate with the British Raj in subduing the Nagas. The British never contemplated anything more than the exercise of a general political control over the hill tribes, and, if necessary, the establishment of a military post to overcome the ill disposed while protecting the peaceful ones. Anything beyond this was not desired. However, the colonial machinery sanctioned Mr. Grange's deputation to receive the submission of the Angami Nagas and to choose a post easily communicable with Assam or Cachar. The authority was at liberty to place a permanent garrison under a native officer for protection of British subjects and encouragement of communication.

The policy of non-interference with the internal management was to be followed, and near this garrison a market was to be established for enhancing trade and commerce (Political Proceedings, 1839; and Political Proceedings, 1840). A letter from Captain Jenkins reported that he had given instruction to Mr. Grange to subdue all the Angamis north of the Water-pent by inviting Manipur to subdue all the area south of it. This subjugation was, he explained, only to be carried far enough to stop outrageous tribal raids. From these tribal chiefs hostages for good conduct were to be demanded. All that the British sought was peace and free intercourse; those who acknowledged their virtual supremacy and gave small tributes were to be admitted to terms (Political Proceedings, 1840). It can be concluded that the British wanted peace and tranquillity for their larger imperial expansion. Having invited the local militia and native rulers to subdue the frontier tribes they successfully sustained colonial interests in an environment where most of the native rulers were subjugated.

Geographical Linkages and Implications

The potential of North East India's geographical linkages with the neighbouring countries has never been translated into actual reality. India's liberation did not have a liberating impact on the relationship between the North East region and South East Asia. The North East frontier region became an isolated landlocked territory representing as much as 7.97 per cent of the Indian geographical area. It has Bhutan, Bangladesh, Burma, China and Tibet as her immediate neighbours with a tenuous link to India. Its natural orientation links it towards north to Tibet and China and east to South East Asia. On the contrary, mainstream development debates argued that the main stumbling block for the region's economic development is its disadvantageous geographical situation (Sinha, 1994, 13).

There were three routes from Assam to Burma in early times: one by the valley of the Brahmaputra upto the Patkai range to Upper Burma, the second through Manipur to the Chindwin valley, and the third through Arakan to the Irrawaddy valley. All these routes met on the frontier of Bhamo and then proceeded over the mountains and across the river valleys of Yunnanfo, i.e., Kunneing, a city in China's southern province. Manipur's Royal Chronicle (*Chietharol Kumbhaba*) record trade between Manipur and Yunan province of China in the year 1630. Manipuris learnt the art of manufacturing gunpowder from Chinese merchants who visited the kingdom. Trade continued as late as 1813 A.D. Influence of Chinese trade and technology was felt in Manipur, namely, in the introduction of silk and silkworm rearing, brick making and making of gunpowder. Indians also proceeded to Far East by land route through Bengal, Manipur and Assam. They reached lower Burma through Arakan and upper Burma through ravine passes in the Patkai range or Manipur hills. Two trade routes with China were along the Irrawaddy and the Salween rivers, and another one down the

Chindwin River. Through Manipur, the traders journeyed to Afghanistan where Chinese silk was exchanged for European gold. The traders travelled in caravans and covered the journey in three months time (Kabui, 1991, 13–14).

Trade between Assam and China got an abrupt boost with the opening of the Stilwell Road. The historical background of this road can be placed under three heads :

1. The Ancient,
2. The Ahom, and
3. The pre-Second World War.

It practically followed the course of the ancient route that links the Upper Irrawaddy and the Chindwin region to Assam. Despite the lack of evidence on the trace of many of these linkages, scholars believe in the existence of a very ancient route that connected the Brahmaputra valley with Yunan. It is thought to be a part of the ancient Southern Silk Route that linked southern China with India. Entry of Ahoms to Assam in early thirteenth century was through this route. Diplomatic missions, small trade, movement of people and even invading armies followed this route during the Ahom period. The Burmese army followed this route in their invasions of Assam during 1817, 1819, and 1821. Between 1828 and 1885, the route was surveyed and explored by several British officials and even a proposal to construct railway line in this route was mooted. The Irrawaddy and other land routes immensely reduced importance of the ancient Patkai-Myitkyina route when the British got direct access to the whole of Upper Burma and Shan States bordering China's southern frontier from Rangoon. Nothing was done for its development by the colonial government. Ultimately it turned into a purely local route and smugglers' track. Reopening of the Stilwell Road involves the opening of a small portion of both sides of the border. It is, therefore, primarily a matter of policy decision on the part of the governments of both countries (Nath, 2004, 39).

Many scholars have rightly pointed out that the international borders shared by the North-Eastern region have been artificially created. The outcome has been the elimination of the region's trade, commerce and other linkages that existed in pre-partition days. Using the region's two per cent perimeter as a major linkage point with the rest of India, and at the same time checking the inflow of goods and people from across the remaining 98 per cent has been a gigantic and counter-productive task (Sachdeva, 2000).

The region's proximity to the dynamic South East and East Asian economies is keenly noted. Economic integration of the region with its transnational neighbours is mooted as an avenue of new opportunities. Following the 1947 partition the region lost its connectivity and market access. Colonial geopolitics had cut the region off from its neighbourhood across the eastern and northern borders. British colonial decisions to draw lines between the hills and the plains, to put barriers on trade between Bhutan and Assam and to treat Burma as a buffer against French Indochina and China severed the region from its traditional trade routes—the southern trails of the Silk Road. Moreover, construction of the region as a frontier by the colonial raj and subsequent creation of imperial stratifications brought structural changes in the internal symbiotic ties between the hills and the valleys. While colonial rulers built railways and roads mostly to take tea, coal, oil and other resources out of the region, disruption of its old trade routes remained colonialism's most enduring negative legacy. After a century and half, an opportunity to undo the effects of colonial geopolitics is emerging. However, there is growing apprehension that there may be more rhetoric than substance to the hype of the region becoming India's gateway to South-East Asia (Baruah, 2005). It is also crucial to assess whether there exists any basis for trade in terms of compatibility between the resource structures of the region and the demand structure of the neighbouring countries (Thiyam, 2005).

Despite the optimism regarding integration of the region's economy with dynamic South East Asia, the '(in)security' phobia looms large. It has been argued that the prism of security through which Indian policy-makers view the North East is likely to remain a major factor inhibiting serious investment of resources—intellectual, political and material—to make the region India's gateway to South East Asia. Despite India's growing official ties with Myanmar, the country's uncertain political future will remain a stumbling block in pursuing policies with a longer time horizon (Baruah, 2005). India's policy dilemma continues, as the region's age-old political problem looms large in the midst of attempts to reconcile the demands of a globalizing economy that relies on greater opening. The current situation not only represents a failure of the economic policy framework in the region, but also a weakness of the country's foreign policy in ignoring South East Asia for a long time. Arguments are made against the current conventional wisdom of spending more money on development in the hope that it will get the region out of its current predicament and concludes by outlining a bold vision for the region's future—political integration with India and economic integration with South East Asia (Ramesh, 2005). Synchronization of Indian trading policies and security obsession may lead to different political problems. With a well thoughtout long-term policy, this region has the potential to emerge as a strategic base to tap the potential of contiguous markets of China, Myanmar, Bangladesh, Laos, Thailand, Vietnam, Cambodia as well as Malaysia, Indonesia and beyond.

Impediments to Development

Partition transformed the region—once at the crossroads of South and South-East Asia—into a landlocked outpost of a large continental economy. The huge landmass comprising the seven States (Assam, Arunachal, Manipur, Meghalaya, Mizoram, Nagaland and Tripura), approximately 225,000 sq. km., became a landlocked hinterland and lost its natural advantage because of trade and industrial policies pursued by independent India (Khanna, 2005). It has been relegated as the remote landlocked and backward region of a dynamic economy. Structure of the region's economy resembles the economy of the least developed parts of the globe; the primary sector accounting for 55–60 percent of the income with heavily underdeveloped secondary and tertiary sectors. The secondary sector constitutes a meagre 11–16 percent of the total income, making it India's industrially most backward region (India, 2004). During the last fifty odd years the general approach to the problems of the region—the growth path and strategy adopted—seem to have been dissonant from the needs and aspirations of the people lacking proper appreciation of ground realities and their real developmental needs. There have been *ad hoc* inputs to situations as they emerge with no indepth study of the problems and planning perspectives. Years of centrally planned developmental efforts have failed to keep pace with the aspirations of the people and address their basic needs. Disturbingly close to thirty per cent of the population still remain below poverty level. NSDP growth rates remain somewhere around four percent and even this has not led to shared prosperity; it has benefited only about 13 percent of the population.

India's inward looking economic strategy deprived the North Eastern States of their natural markets and access to products produced in the neighbouring countries across the border. All products consumed in the region are imported from distant manufacturing regions of India. A World Bank study estimated that costs of logistics and damages added 60 percent to the cost of a bag of cement and 14 percent to general cargo when moved from Calcutta to the North East. Isolation and lack of economic opportunities transformed the North East into a dependent region. Heavily militarized, it remains at the receiving end of the Indian military might. Economic engagements with the neighbouring countries came to be based on the strategic and military postures of New Delhi. Successive regimes in New Delhi have been unable to assess the consequences of their isolationist policies curtailing social and economic links of the region with its neighbours. While other parts of the country moved towards massive industrialization and technologically driven agriculture, the region was left to the military and police to manage the only development objective seemed to be the building of military infrastructure to secure the 'frontier'.

Dramatic expansion of trade with India's eastern neighbours, which constitute a quarter of its total trade, has no impact on the North East since most of this trade expansion takes place through seaports. The Indo-Myanmar Border Trade launched in April, 1995 with the opening of the Tamu (Myanmar)-Moreh (Manipur) sector proved a colossal failure. There are large numbers of non-trade barriers and transit restrictions curtailing possibilities of rapid expansion of trade (Khanna, 2005). The most significant of the non-tariff barriers was the restrictions on transit, visas and customs regulations. Local content requirements (stipulation of approximately 30 percent value addition in domestic market) and quarantine regulations on agriculture commodities meant that the so-called Border Trade Agreement was a smoke screen behind which India protected its market from even its weak neighbours.

Smooth trade across Indo-Myanmar border will be of advantage mainly in reduced costs while accessing the markets of South East and even East Asian countries. Despite this potential, trade through the Manipur-Myanmar route has remained small and insignificant amounting to a few crore per year with little impact on the regional economy. Meanwhile, trade has been stagnant and there seems to be differences with Bangladesh over transit arrangements that India seeks for its links to the North East. Tibet is still closed in the absence of a border agreement. Above all, links to Yunan through Manipur, Mizoram or Myanmar are not on the cards. In all probability the bulk of trade with the Greater Mekong Sub-region (GMS), Bangladesh and Association of South East Asian Nations (ASEAN) is likely to move through the international sea-lanes, completely bypassing the North East. The regions with substantial gains so far are Chennai, Vizag and the Calcutta port on the eastern flank of India. It would be reasonable to argue that given the state of infrastructure and the poor state of road, rail and air links with the neighbouring countries in the North East, the bulk of the trade is likely to move through the sea ports of India (Intercultural Resources and Forum for Indigenous Perspectives and Action, 2006). These schemes are not meant for regional development as publicised but 'for investment and trade liberalization' which result in extensive exploitation and expropriation of natural resources such as water, forest, fisheries etc. severely affecting livelihood and environment. Projects are based on potential profits for investors and not on long-term economic interests of the local people. Ironically, not even the local representatives of the region are aware of any kind of policies and development plans indicating the absence of participation in development policy making.

Delineating the Development Paradigm : Its Implications

While discussing the problem of development of the North East, a study of the region's balance sheet is important. On the credit side the region has the following (*Yojana*, December 2006, 14):

(a) Abundant natural resources and tremendous potential in the growth of an agro-horti-forestry sector;
(b) Substantial mineral deposit (in Assam and Meghalaya);
(c) Unique bio-diversity;
(d) Vast water resources including hydel power potential with more than 50,000 MW identified capacity;
(e) Great promise for tourism development;
(f) Proximity to one of the world's fastest growing economies (South East Asia);
(g) A literate population;
(h) Rich heritage of handloom and handicrafts; and
(i) A democratic traditional system of local self government with community spirit permeating the entire social system.

On the debit side are the following :

(a) Inadequate basic developmental infrastructure;
(b) Geographical isolation that reduces mobility;
(c) High rainfall and recurring flood in the Brahmaputra valley;
(d) Lack of capital formation and proper entrepreneurial climate;
(e) Slow spread of technology, and
(f) Absence of a supporting market structure and adequate institutional finance structure.

An overall assessment of development issues necessarily underscores the need to question whether opening up of the region can solve the problems inflicting the region. Dilemma on the 'security' issue may contradict well-planned economic opening up of the region. It has been rigorously argued that the 'security' curtain under which the Indian government has tried to keep the North East 'secure' is in the same line pursued by the British in the region. New Delhi has adopted the same 'colonial' policy of reducing the North East to a mere buffer zone since British departure from the subcontinent (Egreteau, 2006). Upsetting of geostrategic configurations after the partition in 1947 followed by the political instability and civil war in independent Burma from 1948 isolated the North East. An important issue to be addressed revolves round the need for resettlement of social spatial settings of the region in order to make the environment more conducive for economic development before opening up the 'frontier'. Above all, another concern pertains to the region's requirement for infrastructural development for making it ready for productive economic intervention.

It is crucial to study the policy dimensions and implication on the region's polity, economy, society and culture. Politics of defining the region in developmental terms involves the issue of New Delhi's commitment towards the States in the region to take their own fast track route to collaboration with the neighbouring countries. The problem of overlapping jurisdictions over sub-regional or zonal grouping comes up (Sobhan, 1999). Territorial transformation of the North East, which was engineered to suit the growing interests of the British in the region as part of new defence system of the British Empire persists even today.

REFERENCES

Baruah, Sanjib. 2005. The problem. Paper presented in Gateway to the East, a symposium on North East India and the Look East Policy, June 11–13, in Guwahati, India.

Egreteau, Renaud. 2006. *Instability at the Gate: India's Troubled Northeast and its External Connections*. New Delhi: Publication of the French Institute in India.

Government of India. 2005. *India 2004*. New Delhi: Ministry of Information and Broadcasting.

Intercultural Resources and Forum for Indigenous Perspectives & Action. 2006. *Insidious Financial Intrusions in India's North East*. Imphal.

Kabui, Gangmumei. 1991. *History of Manipur*, Vol. 1. New Delhi: National Publishing House.

Khanna, Sushil. 2005. Economic opportunities or continuing stagnation. <http://www.india-seminar.com/>

Mackenzie, Alexander. 1884. *History of the Relations of the Government with the Hill Tribes of the North East Frontier of Bengal*. Calcutta: Superintendent of Government Printing.

Nath, D. ed. 2004. *Reopening of the Stilwell Road: Prospect and Problems*. Kolkata: Anamika Publishers and Distributors.

Political Proceedings, 2 October, 1839, Nos. 89–90.

Political Proceedings, 1 January 1840, Nos. 112–13.

Ramesh, Jairam. 2005. Northeast India in a new Asia. <www. http://www.india-seminar.com/>

Sachdeva, Gulshan. 2000. *Economy of the North East: Policy, Present Condition and Future Possibilities*. New Delhi: Konark.

Sinha, A.C. 1994. *North Eastern Frontier of India: Structural Imperatives and Aspects of Change*. New Delhi: Mittal Publications.

Sobhan, Rehman. 1999. *Transforming Eastern South Asia: Building Growth Zones for Economic Cooperation*. Dacca: The University Press.

Syiemlieh, David R. eds. 2006. *Challenges of Development in North East India*. New Delhi: Regency Publications.

Thiyam, Bharat. 2005. India's Border Trade with its Neighbouring Countries: With Special reference to Myanmar. *Margin: The Journal of Applied Economic Research* 1, 4: 359–382.

4

In the Confines of Geo-Spatial and Extra-Economic Identity : Development of the North East India

Arambam Noni

Prologue : Tracing Development in the Mist of History

Historical perception on the North East continues to arrest development as well as economic growth of the region perpetually. This is especially true when seen in the context of the post-cold war realities where regional and bilateral trading mainly in distant borders of countries have grown manifold. The question of development remains in the confines of the State-centric mindset of the national governments that view North East Region (NER) as inaccessible, distant and disloyal converting the region simply into a buffer zone. To elaborate the argument further, it is worthwhile to mention some of the 'generally' accepted or commonly held views on the North East and consequent particularization of the region in terms of imagination and associated ritual like practices. For example, one of the regular topics that any high school student of economics in India tries to memorize in their first encounter with the North East is the idea that their textbooks enumerate. These textbooks cite the region as a distant frontier, landlocked, inaccessible and economically backward, and that attempts to develop the region remain unviable. Mainstream perceptions grow up persistently getting informed that the underperformance of the region is something inherent in its geo-spatial settings.

Perpetuation and naturalization of this image has produced a fixed identity of territory and people of the region. Such a confined particularity has resulted into a lingering constitution of a lifetime imagination on perspectives, practices and identities whereby the modern North East appears to be a 'politico-cultural and economic other'. Spacing of North East as a 'Region' in the 'national-terrain' of post-British India spells out the paradoxes of continued inheritance of a colonial articulation typical of many hurriedly enforced 'nation-ness'. This historiography is symbolic of post-colonial India's idea of its constitutive frontiers as it continues to connote two physical bodies one, an 'area' and second, a 'boundary line' (Thingnam, 2006). Such a perception is based on recurring effects of colonial mapping of possessed territory and its people through manipulative cartographic idioms to claim suzerainty over the possessed areas that consequently draw a boundary line by validating through imaginary crafting of maps even over the inaccessible distant territories (Barrow, 2003).

Imperial perceptions of the North East were well co-opted in the newly acquired 'post-British' Indian nationalist ideology. Colonial pattern of exploring 'new terrains' was embedded in many imperial projects like military expeditions and anthropological researches whose subsequent conceptualization and practices produced dissenting historical companions in the past and present as well. British colonial encounters with the frontier produced a rudimentary outline of political and cultural mapping in which 'otherness' of the North East was a reality. Continuation of conceiving the region as the 'other' was evident in the terms of reference made by Sardar Vallabhbhai Patel, the First Home Minister of independent India in a letter written to Jawaharlal Nehru in 1950. Patel said, 'The North-Eastern areas are by no means close and intimate to mainland India. The people inhabiting these areas have no established loyalty or devotion to India and are not free from pro-mongoloid prejudices' (quoted in Subba, 1996, 97).

The historical statement cited above is a strong case in point that uncovers many undercurrent perceptions about the region in the nationalist mindset and the stories of integration that evolved during India's integration politics corresponds to the inherent dilemmas of the 'post-colonial' in India. Nature of 'integrationist' politics pursued by India in the region completely thwarted newly attained representative democracies in the region. For example, in 1948, Manipur became the first geo-political entity in the entire subcontinent to hold an independent election based on adult franchise. However, this celebration of democracy proved to be short-lived, as it soon became a victim of 'coercive national practice'. To detail such a practice, in 1949, V.P. Menon, a senior representative of the Government of India, invited the King of Manipur to a meeting in Shillong on the pretext of discussing the allegedly deteriorating law and order situation in Manipur. Consequently, the King was made to sign the Merger Agreement 'under duress' on September 21, 1949 at Shillong (Nag, 2002, 204). The letters written by the Maharaja while he was under house arrest in Shillong reflect what actually happened in the making of India. One of the letters dated September 18, 1949 reads :

> Now that the sovereignty of the State has been vested in the people, it would be in the fitness of things to hear the people's voice and learn their sentiments so that the line of action may not in any case be unconstitutional (*Resistance*, 1975, 52).

The Maharaja further wrote in another letter:

> [. . . . . .] I consider that I have a guardian in the person of your Excellency who guides and helps me in my endeavour towards ameliorating the welfare of my people (Manipur). The only thing is that this being a great change it should be done constitutionally and democratically with an eye to the will of the people whose co-operation I confidently hope we will secure (Nag, 2002, 204).

In a response to the discomfort of the Raja, Vallabhbhai Patel asked the then Governor of Assam, Sri Prakasa, 'whether he (the Governor) did not have a Brigadier in Shillong' (Rustomji, 1973, 109). The non-consented consensus authored in Shillong marked the entry of independent India into Manipur as the Indian nationalists' authoring of its dominant self in Manipur was pursued as if invented artifacts were an adequate form and contents for making a 'nation'. In other words, the process of Indian 'integration' began with an 'authoritarian accent' (Baruah 2001, 12) thereby producing a dissonant 'historical time'. Thus, introduction of Indian national politics in the North East amplifies a historical absence of India in the region whose 'official nationalism' inherits the features of an imperial body where the presence is marked by a continuous absence of legitimacy (Meetei, 2006, 242). The monologue of the post-colonial Indian State discourse in elevating the differences of identities to a common subjecthood within a hegemonic space is a source of dissent, resistance and protest from the margins (Swain, 1996). This imagining of 'nationhood' affirms a historical consolidation of opposing simultaneous historical currents that constructs its own self against the very project of dominant nationhood constituting historical 'national aspirants' in its dissenting form we find, today, constantly reproducing dissents to the 'idea of India' (Meetei, 2006, 240). It is this nature of dissented intersections of historical identities which has given causal grounds to the contemporary shades of movements on the grounds of 'historically distinct, denied, coerced, simultaneous and independent other'.

Deeper investigation into this aspect can assist in understanding the contemporary nature and framework of 'official authoring' of people and region. Similar endeavour can elucidate the nature of polemics inherent in the polity, economy and military imaginations augmented *vis-à-vis* the region in the past as well as the present. To pose this proposition thoroughly, the perspectives and polemics (extra-economic rationales) associated with the politics of development and planning in India can be examined.

The Politics of Planning

Practices in the newly acquired India's independence had two paradoxes—one, economy was articulated through an economic critique of colonialism as a necessary legitimising condition of the nation thereby putting a case for 'national' development. And, second, the State was connected to the people-nation through a procedural form of representative government. The paradox lay in the disconnection between these two projects. Similarly, an inherent connection among the contents and practices of the 'nation, people and development' in the post-colonial India was largely absent as it was mainly based on modalities of how to use its sovereign power. Thus, what the people could express through the representative mechanism as their 'will' was not necessarily what was good for their economic well-being. This disconnection largely produced contradictory implications on state policies. Development of the North East is one such domain that has experienced similar asymmetry. Also, this emanated on account of the post-colonial idea of development, which implies a linear path, directed towards a goal premised upon a rational consciousness and will of the 'national Indian state'. This resulted in perceiving 'development' as a process to impact upon the whole of society premised upon one consciousness and will that represents the interest of rationally ignorant 'whole' which potentially attempt to put the interests of the 'particular' as something to be always subsumed within the 'whole' (Chatterjee 1997, 277).

Historicity of the 'whole' and 'particular' has been employed continuously in confining the NER and its economy in distinctive features of geo-spatiality and regionalism. In other words, a post-colonial state like India has been employing geo-strategic and economic rationale as tools of mapping the peripheries such as the NER. To cite a case, in addition to its economic interests, a testimony to this 'geo-strategic rationale' could be well seen in the context under which the entry of British Empire into today's North East was conceded. For instance, it was largely to counter the French prevalence in Burma. Over the years the conceptualization of the NER as a politico-military space has produced many categories of distinctiveness where the idea of spatiality remains a ground for racial differentiation, cultural split, political contradiction, economic deprivation and dependence. The mixture of these conflicting dynamics is well carried out in post-colonial practices of governing the frontiers. The case of NER is explanatory to this proposition. A critical analysis of the nature of development planning and contemporary policies of economic 'openness' across the previously 'hard frontiers' can serve as a strong area of reviewing post-colonial India's idea of North East.

Economic planning in India since independence is fairly subjected to regional disparities in addition to the overall poor economic health of India. The economy inherited from imperial times did add to the mounting challenges to the planners as it was highly disintegrated. Precisely on account of these reasons, the First Five Year Plan of India tried to stabilise and rehabilitate the economy from the colonial devastation of 1947. The first batch of Indian planners did know that a wider diffusion of industrialization was desirable for balanced economic empowerment of the country which was only possible with due attention to the development of those States and regions which were backward though subsequent phases of Indian Planning and implementation did not convert these concerns into reality. For example, in a highly agrarian country like India, the simultaneous emphasis on heavy industrialization was always an uphill task, which on the one hand did affect the smooth essentialization of agrarian production system.

Indian planners and policy makers in their hurried attempt to attain economic growth largely emphasised on industrialization that required intensive labour, capital and consumption markets. For this, regions like the North East did not figure anywhere in the actual economic imagination of the planners. As a result, NER soon transformed from being a potential trading power centre to a small 'captive market' for goods of all kinds produced from mainland and non-regularized neighbouring markets as well. For instance, its tea and forest resources, an important source of export earnings initially for India, soon lost their importance as Indian national exports diversified towards manufacturing sector. This tragedy for the economies in NER was further seen in the Third and Fourth Five Year Plans. The perspectives of Third Five Year Plan endorses that there are appropriate possibilities of regional economies like North East when it said :

There are examples of regional development...there are other industries whose possibilities need to be fully explored such as labour intensive industries of the traditional type, small scale industries of modern type, agricultural processing and forest industries...which are suited to its conditions...' (Government of India, 1961).

But the practice of appropriate possibilities for 'distinct economies' (North East) did not find its due place in the subsequent politics of planning in India. This can be seen in India's overt industrialization strategy during 1956–1991 based largely on import substitution and inclination towards exports thereby resulting in the high tariff walls. For a region like the North East where 98 percent of its territory is part of the international borders, access to the international market at low tariff is a pre-condition for a vibrant possibility of procuring commodities both meant for consumption and business. Ironically, most of daily consumer items in the North East come from distant manufacturing regions in India. For example, construction materials like cement are procured from Orissa and Madhya Pradesh while chemicals come from Gujarat and Maharashtra which is not cost effective due to the added transportation cost. An estimate by the World Bank study states that costs of logistics and damages adds 60 percent to the actual cost of a bag of cement and 14 percent to general cargo, when moved from Calcutta to the North East (World Bank, 2001). Ironically, the 'official' documents have even stated that the structure of the economy in the region resembles the economy of the least developed parts of the globe, with the primary sector accounting for 55–60 percent of the income, and underdeveloped secondary and tertiary sectors. The secondary sector constitutes only a meagre 11–16 percent of the total income, making it India's industrially most backward region (India, 2004).

Economic underdevelopment of the region is to be investigated to find its causal explanations not only based on the exclusive consequence of the long neglect but also the historicity of misappropriating the region with its strong aftershocks. It can be reasonably argued that the failure of economic development in the North East is largely because of an inappropriate economic policy framework which has created an unbalanced and unsustainable economy destroying the basis for institutions of a modern market economy in the region (Sachdeva, 2000). The long inappropriate economic practices have produced high levels of poverty ranging between 42–58 percent making it one of the most backward areas in India. Its per capita income, too, is far below the national average, with Assam having a per capita income of Rs. 10,000 in 2001–02 compared to Rs. 18,000 for India (Khanna, 2005). Therefore, the human development indicators (HDI), too, are equally dismal. In terms of infant mortality, the States of the North East are ranked far below the national average.Though, States such as Arunachal and Tripura have child mortality rates above 100 (NCAER, 2004) despite the high rates of literacy in the region. Having identified such a bleak economic experience, the prospects of opening up of the region mooted recently as a policy framework need to be critically reviewed in the light of the historicity and geo-spatial identity of the region.

Look East Policy : Uncovering the Extra-Economics

The above attempt to engage with the 'blemish perspectives' of perceiving and practice of political economy in the NER is well reflected in the consequent underperformance. From the initial years of phased planning to the present day opening of markets of the 'previously closed' NER, actual economic growth and empowerment of the people in the region have not happened whose causal explanations lie in the supplanting of economics by extra-economic rationales. The extra-economics of the economy is linked to the nature of Indian nationalist imagination that perceives NER with 'caution and suspect'. The intricate dimensions of the geo-spatial regimentation of NER in conceiving the present Look East Policy (LEP) corresponds to larger politico-economic rationale of the Indian State. The seven States, Arunachal, Assam, Manipur, Meghalaya, Mizoram, Nagaland and Tripura, which originally constitute the NER is linked to the mainland only through a thin corridor of 20 km. The region has historical experiences of being distinct but its marriage with the nascent 'post-colonial' India did not result in the mellowing down of its distinctiveness in the 'mainstream psyche' and the common psyche of the people in the region as well. Continuity of the civil service, police administration, judicial system including codes of civil and criminal law and the armed forces as they existed in the colonial period. Operationalization of the post-colonial is again aptly expressed in the NER when post-independence laws like the Armed Forces (Special Powers) Act, 1958 finds its original form in a colonial Ordinance of 1942 that was enacted to suppress 'Quit India Movement' against the Imperial British. In the case of the region, the distinctiveness of this 'otherness' gets reconfirmed in the everyday life experiences marked by violent encounter between the State and people (manifest in every aspect of socio-economic and political articulation) in the region.

Debates over the LEP, despite its economic centrality *vis-à-vis* the region, need to assess the extra-economic issues much more powerfully. This is a hint to the larger historical experiences of the region as a frontier location, economic and political body. To recapitulate the emergence of LEP, it was actually in 1991 LEP was conceived with the acceptance of structural adjustment programmes (SAP). Opening of 'post-colonial' India towards South East Asia including China resulted in dismantling trade barriers, abolition of import licensing so as to promote inflow of global capital. India's growing ties with the 'East Economies' effected over the years through its Preferential Trade Arrangements (PTAs) with the neighbouring countries, even by going outside the South Asian Preferential Trade Agreement (SAPTA) or South Asian Association for Regional Cooperation (SAARC) framework and entering into bilateral and multilateral trade agreements like South Asia Growth Quadrangle (SAGQ), South Asian Sub-region for Economic Cooperation (SASEC), Bangladesh-India-Myanmar-Sri Lanka-Thailand-Economic Cooperation (BIMSTEC), etc. has had little or virtually no impact on the North East. LEP has made substantial progress in terms of India's trade with other developing countries of Asia. The share has doubled contributing about 30 percent of India's trade, while South East Asia's share has doubled to

24.2 percent. In other words, about a quarter of India foreign trade now comes from its Asian neighbours. India's trade with countries bordering the North East has seen the most dramatic expansion share going up more than five times (from 1.7 percent to 8 percent). Most of this trade expansion has taken place through the seaports (Khanna, 2005). At the same time, opening of the region to various big private operators for involvement in forest management and environmental control, on water resource management and electricity are likely to deprive the right of the commoners on habitable/cultivable land, water and essential services.

The 'dollar syndrome' created by LEP generate doubts over its role as a viable economic alternative to the long economic backwardness of the North East. LEP is devised to attract global capital on the pretext of economic development that potentially converts the region into a transit corridor without any substantial material and infrastructural empowerment of the indigenous communities (Kangujam, 2006). It is this contextualization of LEP that apparently hints to the obvious continued imprisonment of the region's economy to its colonial experience in the past. During the British colonial period roads and railways were mostly built to extract tea, coal, oil and other forest resources out of Assam, whereas other parts of the region largely remained more of a passage to secure and further economic and political expansion of the imperial power. Imperial British's entry into the region was also largely strategic in its bid to contain French expansion in Myanmar. This is a pointer to the predicaments of LEP as most of the bilateral and multi-lateral trade agreements with South East Asia have produced an odd mixture of economics and politics. Overt presence of the State in terms of its inability to recede *vis-à-vis* the globalization trends noticed elsewhere indicates serious economic distortions. Obsessive security concerns in the region downplay the economic potentials and prospects as well.

Economy of the region gets shaped by 'security' concerns on the one hand and results in overt presence of the State with constant militarization of civil spaces including economic planning. Actual trading, economic development and democratic aspirations in the region are easily strangulated in the face of India's pursuit for 'security'. For instance, India's foreign policies with regard to Bangladesh, Myanmar and Nepal have not seen much of committed engagement at the level of deepening democracy in the region. India's position on the question of democratization in Myanmar and Bhutan reflects a deficit in actual democracy. Not surprisingly, every regional economic cooperation accord signed there are corresponding technological and military accords. This bizarre mix of economics and militaristic politics is the manifestation of a peculiar mode of political conception of India confirms that it cannot escape its externality featured in the politico-economic imagination of the 'mainland'. For instance, BIMSTEC, 1997, was proposed by India and Thailand as a vehicle for promoting trade and tourism in the Bay of Bengal region. Bhutan and Nepal have been subsequently added. Expansion of BIMSTEC proved to be militarily useful since India has started receiving military cooperation from countries like Bhutan to tackle armed movements in the NER.

In 2000, further growth in the trans-regional cooperation has been seen again when India floated the Mekong-Ganga Cooperation (MGC) initiative with countries such as Myanmar, Thailand, Laos, Cambodia and Vietnam for regional cooperation in tourism, cultural exchange and education. But the prospect of NER in the MGC is bleak as there is strong possibility of shifting to maritime channels. One of the causal explanations is the fact that policy perspectives are suspicious not only of the region's economy, but also of ethnic and social ties with the neighbours (Khanna, 2005). India's recent extension of its economic ties to Vietnam and Indonesia under the LEP vision can be regarded as a crucial move to counter the alleged Chinese encroachment into what India perceives as its domain, the Indian Ocean and Myanmar. Thus, there is manifested requirement for the Indian security activities to reach upto the eastern range of Indian Ocean so as to influence maritime reach from the Bay of Bengal to the South China Sea. Since the beginning of 1990s China has allegedly helped Myanmar in construction of intelligence sites and naval bases in Great Coco Island, which is uncomfortable for India as it is close to its military installations in the Andaman

Islands. Thus, the region has become a theatre of contesting powers as it has ceaselessly seen China's continuous interests in Myanmar,including Arunachal Pradesh, with counter attempts from India. Chinese technicians also built a series of smaller ELINT/ SIGINT, satellite stations, along the Burmese coast of Bay of Bengal. As a result, it has led to achieving a thorough and overlapping coverage of the Bay of Bengal and Strait of Malacca. Result of this military contestation is reflected in the defence deals that India signed in 2001 with Indonesia and in 2000 with Vietnam. As a prelude to economic engagements, military politics continues to assume centrality in the region.

Conclusion

The cartographic regimentation of the NER has remained a major factor inhibiting the investment of resources—intellectual, political and material—to make the region India's gateway to South East Asia. Opening up of economies with sufficient regional protection in terms of economic interests like proper infrastructure, resource and manpower utilization by not overriding the cultural and political questions of the communities in the region may ensure some substance to the logic of LEP. Nevertheless, the paradox for India is to set itself with the demands of a globalising economy on the one hand and elevation of armed struggle in the region on the other. Unless a policy rethinking on the excesses of extra-economic reasoning is done with a basic commitment towards deepening democracy, a sustainable vision of trade in the region can never emerge.

Today, the danger of border trade is its tendency to indulge in trade actually without involving the concerned region whereby it just becomes a transit point. The nature of economic options available in the LEP do not seem to be replaced gradually by the goods of local origin in order to establish the organic link between local resource mobilization, production and trade which will ensure development of the bordering region. An intention of receiving global capital and using the region as a frontier military check post to the contestations within and restraining the influence of the rival powers in the region on the pretext of development is bound to have recurring backlashes.

REFERENCES

Meetei, Arambam Noni. 2006. Multiple Nationalisms in Manipur: A historical time and its reproduction. *Eastern Quarterly* 3, IV. January–March: 240–50.

Barrow, Ian J. 2003. *Making History: Drawing Territory*. New Delhi: Oxford University Press.

Baruah, Sanjib. 2001. General as the parallel political system. *Himal* 14, 6 June: 12–20.

——. 2004. Between South and South East Asia North East India and Look East Policy. CENISEAS. Paper 4, Guwahati.

Chatterjee, Partha. 1997. Development planning and the Indian state. In *State and Politics in India*, ed. Partha Chatterjee, xyz pp. New Delhi: Oxford University Press.

Government of India. 1952. First Five Years Plan. New Delhi: Planning Commission.

——. 1961. Third Five Years Plan. New Delhi: Planning Commission.

——. 1998. Transforming the North East: High Level Committee Report. New Delhi: Planning Commission.

——. *India* 2004. New Delhi: Ministry of Information and Broadcasting.

Kangujam, Sanatomba. 2006. Trade, security and strategic concerns: India's Look East Policy and the suspect politics of ambivalence. *Alternative Frames* I, III. April–June: 57–79.

Khanna, Sushil. 2002. *Trade and investment in South Asia Sub-Region: Barriers and Opportunities*. Calcutta: IIM Mimeo.

——. 2005. Economic opportunities or continuing stagnation. Paper presented at a seminar on Development of North East, June 19–22, in New Delhi, India.

Nag, Sajal. 2002. *Contesting Marginality; Ethnicity, Insurgency and Subnationalism in North-East India*. New Delhi: Manohar.

NCAER. 2004. *East India: Human Development Report*. New Delhi: New Delhi: Oxford University Press.

Resistance IV, 3. 1979. Colonial History of Manipur: Betrayal in Shillong. September 25: 52–3.

Rustomji, Nari. 1973. *The Enchanted Frontiers*. New Delhi: Oxford University Press.

Sachdeva, Gulshan. 2000. *Economy of the North East: Policy, Present Condition and Future Possibilities*. Delhi: Konark Publications.

Subba, T.B. 1996. Ethnicity, culture and nationalism in North-East: A conspectus. In *Ethnicity, Culture and Nationalism in North East India*, ed. M.M. Agrawal, 38–48. Indus: New Delhi.

Swain, Nirmal Kumar. 1996. The post-colonial Indian state: A paradigm for emergence of new identities. In *Ethnicity, Culture and Nationalism in North East India*, ed. M.M. Agrawal, 77–83. Indus: New Delhi.

Thingnam, Sanjeev. 2006. Redefining frontier through LEP: A colonial articulation of Manipur. *Alternative Frames* I, III. April–June: 4–19.

World Bank. 2001. *Forging Sub-regional Links in Transportation and Logistics in South Asia*. Washington DC: World Bank.

5

State and Development in the North East

Kangujam Sanatomba

One of the questions that development critics so often puts across —and of course, for which much has been endeavoured—centres on the need to rethink development theory in recent times. It is crucial to interrogate whether development should be treated as managerial prescription or an imposition upon those who lack a definitive capacity to take on the challenges of 'under-development'. Prescription as it always involves compelling, to some extent, a change in the identities of those who have been supposed beneficiaries. Who has benefited and in what ways has been targeted in the critiques raised recently. Organizations such as the World Bank and USAID are believed to impose expertise and authority, while silencing alternative voices, promote dependent paths to development, and keep their eyes closed to the economic disparities and power imbalances they create.

Development interventions are intended to move societies from a situation in which they are believed to be worse off, to situations in which they are assumed to be better off. Certainly, there is a great deal of contention on what determines who is 'worse' and who is 'better'. Recently, on the part of development critics, there has been a wide concern to raise social and political dimensions of development. Research on development has become multi-disciplinary, embracing policy analysis. It has started focusing on the major symptom of failed development that is poverty.

Development intervention helps, as it is strongly argued in many of the mainstream argument about promoting development as currently received, to overcome human rights abuses, settle conflicts and violence, bring about peace, protect the environment, and empower women and other minor communities and various cultural groups. While the debate about the pros and cons of different approaches to development still rages, the constant critiques have inspired a quest for diversity among development initiatives in different parts of the world. One such case is the region known as North East India.

Development has been the focus of everybody's concern in the North East and those who have a feeling for bringing about any good change in the region believe that a definitive strategy of ensuring a steady move towards economic development will have immediate ramifications in many aspects of the present turmoil in the region. The issue however is not simply about question of how to bring about development in the region. It is rather about what development is going to do in the North East. Development thereby is loaded with different implications depending on the position one takes according to one's perspectives to socio-economic growth and development in the region. In this context, the whole debate on the role of development in the region falls into two broad themes. First, there are many who believe that economic development will surely lead to an end of the present turmoil in the region. The experiences of disillusion in the region especially among the youth are closely related to the lack of economic advancement to cater to the needs of the present generation and its value system. Secondly, another large section of the divide in the opinion holds that development cannot be the cause of every good change amidst various layers of issues; without first addressing those issues some of which are directly related to lack of development while some others are not related to lack of development. According to this section development can be ensured only when the conditions of its occurrence are primarily addressed and thereby prior sorting out the political/social problems of the region is mandatory.

Apart from these two general perceptions on the role of development, one can look into the consequences, which have been gradually emerging around the whole process of shaping the strategies of development. It can be maintained that one of the consequences or perhaps a fallout of the deeper process involved in the development process in the North East is that it encourages at various levels the psychology of dependency as against collective or even individual initiative. Development strategies in the region to counter the growing tendencies have produced an overtly strong position of the State in its bid to protect its reason to be unchallenged in any of the ways it sanctions. Overwhelming concentration and use of power by the State to endorse development strategies propound a Machiavellian notion and practice of *raison d'etre*. On the one hand, the process of endorsing particular modes of development implants more dependency of State governments on the Centre, and, on the other, dependency of people on the State. Thus current development policies and debates particularly in the official circle always support the statist prescription of development. To put it differently, the raison *d'etre* that supports this particular mode of dependency strengthens the indomitability of State authority. Increasing 'authoritarian' State operation *vis-à-vis* development entails the logic of centre dependent periphery. Development policies initiated and supported by the State in the North East is understandable from an overview of how the region's body politic resembles an entity fed by the Centre. It would be equally reductive to assess the dynamics of dependency in this regard by a facile focus on Centre-State relations or federal equations of the Indian polity without an oblique consideration of the intricate relationship built between development of the North East and *raison d'etre* of the Indian State. Much vaunted debates on Centre-State relations and federal issues have failed to address the complex politico-historical realities of the body politic of the region.

It is generally believed that the region continues to suffer from its geographical position, transport limitations, small market size and low investments. It has also been ascribed to inadequacy of rail, road and air connectivity. However, these limitations are only a manifestation of the nature and kind of development policies framed and implemented over the years. While the region can still be construed in terms of what the policymakers call 'disadvantages', what needs to be considered is that the bottleneck is closely related to particular modes of development promoted under State supervision without proper consultation, participation and involvement of the communities and groups concerned. Hence, despite inherent economic potential, past development initiatives have often resulted in lopsided results. Positive changes on a substantive scale could not be achieved for a complex set of reasons. One of the root causes has been the prescriptive approach, which insidiously ignored the socio-cultural complexity of indigenous population and their interrelations. The precursory rounds of survey given here will substantially offer an understanding that the region is dependent on the Centre while a large sum of money is annually poured down for its development from the Centre.

The Government of India normally gives a total expenditure of almost Rs. 30,000 crore a year for a population of roughly 40 million or thereabout in the region, which amounts to Rs 10,000 rupees per person per year. In Arunachal Pradesh 85 percent of this money comes from the Centre; in Assam 51 percent comes from the Centre; in Meghalaya it is 70 percent; in Manipur 80 percent; in Nagaland 80 percent; in Sikkim central contribution accounts for only about 40 percent; in Tripura 70 percent and in Mizoram 72 percent come from the Centre. In short, of the Rs. 30,000 crore that is spent in North East, year after year, close to Rs. 20,000 crore comes directly from the Centre. Apart from this, the Ministry of Development of North-Eastern Region (DONER), which was set-up as a Department of the Government of India in 2001 with primary mandate to catalise development of the North East, was upgraded to a separate Ministry in 2004. With the establishment of DONER more funds have been channelised with strict regulations on usage. For example, in 2004, DONER Minister P.R. Kyndiah announced a 'special economic package' of Rs. 240 crore to Manipur. The package was meant for projects especially in the fields of education, health, power, transport and communication. DONER Minister Kyndiah said utilization of the funds would be 'strictly monitored... There could be a significant impact on the State's fortunes if our guidelines are adhered to.' Annually the DONER has added to the budget another Rs. 550 crore. Besides the North Eastern Council (NEC) adds another

Rs. 500 crore. This amount in totality is a small part of the enormous amount of funding available to the States through different Central schemes, packages announced by various Prime Ministers from time to time, 'Peace Packages' provided to States like Nagaland and Mizoram etc., grants by international development agencies like the World Bank, and the Asian Development Bank (ADB), which in 2003 approved a master project for the North East of Rs. 2,000 crore (Routray, 2000).

While these are not the least of the whole of the list, since 1990s and particularly from 1996 when Deve Gowda the then Prime Minister of India started a huge jump in the way the Centre gives attention to the region, the trend has been on the rise as different parties at the Centre take the trend rightful in addressing the issues of the North East. Though it is true that New Delhi changes its policy on North East with the change of governments and political parties under whose control the government rules at the Centre and the States, the framework of reference most commonly acceptable to all of them is pumping more funds into the region so that the long time accusation of negligence against New Delhi is countered and somehow development can at least be initiated. It is reflected in the words of the former Prime Minister A.B. Vajpayee in his address at Guwahati in 1998, 'When people in New Delhi think of the North East, they usually think of the geographic distance, which translates itself into mental distance. When people in the North East think of New Delhi, they usually think of the developmental distance, which translates itself into emotional distance.' Prime Minister Deve Gowda in 1996 told that development efforts and counter-insurgency measures must go hand-in-hand in the region and one cannot wait for insurgency to end. He seemed to believe that the situation would improve once the government launches serious development and economic regeneration drives. He gave a Rs. 6,100 crore exclusive economic package for the region. His successor I.K. Gujral further offered an increased amount by Rs. 1000 core, that is Rs. 7,000 crore and then Prime Minister Atal Behari Vajpayee in January 2000 announced a Rs. 10,000 crore special package for the region.

Does one see any substantive improvement in the living conditions, turmoils, experiences and conflicts in the North East as a result of all these efforts? There are many opinions about this question. One which is prominently given and of course shared by many is that inefficient governance, corruption, lack of infrastructure only make the efforts futile and most importantly as it is agreed among the official and non-official circle, aggravates violence and insurgency. The reason, considered very strong in itself, is that insurgency disrupts administration and supports corruption since various armed groups simply feed on money collected through extortion. Consequently government is ineffective and welfare policies cannot simply be worked out. One can look into the statistical figures to discuss this issue. It has been estimated that the per capita income of the North East is much below the all India average. Though the urban poverty ratio in the region has fallen from 36.92 percent in 1973 to 7.4 percent in 1999–2000, the rural poverty ratios continue to be higher than the all India average at 40.40 percent. The population below the poverty line in 2000–01 was 35.13 percent against the all India average of 26.1 percent. This region also represents one of the highest rates of unemployment in India with an unemployment rate close to 12 percent against a total average of 7.7 percent (NER DATA). The most fatal mistake which can be derived from these accounts is that counter insurgency operation and measures associated with it are closely brought together with welfare policies and most seriously the efforts are considered as identical with development strategies. As a result various ramifications emerging out of the very incoherence between development policies and the socio-cultural and political context are neglected while simultaneously blaming the administration that is incapacitated by insurgency for any shortcomings and, therefore, simplistically reduces insurgency to be the root cause of all problems in the North East. So, everything is apparently linked to it. Every aspect of strategies are necessarily linked to insurgency, be it health, poverty and employment or the basic parameters of respectable definition of a goal achieved by any developmental policy. While nothing substantial has been witnessed in the last six decades of State operation in the North East, development still remains the frontline issue—how to bring about development and what to do with it? The invisible mechanism and undercurrents in development dynamics alarm one. Development policies so far have simply been a strategy of enhancing overt *raison d'etre* in the region. Development so far does not do anything apart from serving the State to make its presence obligatory in bid to preserve its legitimacy and so that the rulers may have their positions secure.

One can further consider this point if Assam's case is attended to a little more. Assam is the biggest economy in the region. Dibrugarh, Jorhat, Kamrupa, Nagaon, Sibasagar, Sonitpur, and Tinsukia are districts with high concentration of industries. Among these districts, Kamrupa and Dibrugarh have the highest concentration of industries. Kamrupa district has a large number of large and medium industries. It also has quite a diversified range of industries. Industries in this district are in sectors like petroleum and petroleum products, fertilizer, flour mills, industrial gas, steel and steel products, printing and paper, cement, tubes and pipes, polyester and acrylic, jute, textiles, edible oil and vanaspati, sugar, automobiles, financial companies, plastics, tea, consumer goods, and coal products. Among the dominant industries in this district are steel and steel products (9 units), petroleum related industries (7 units) and consumer goods (13 units). Among the other districts of Assam, Cachar, Darrang, Jorhat, Sibasagar, and Tinsukia all have a number of large and medium enterprises in the tea industry. Tinsukia also has around 19 large and medium enterprises that deal with plywood and timber. North Cachar Hills (N.C. Hills) District has a number of cement plants (Report, DONER).

Since the 1990s growth in per capita income of Assam has almost stagnated. The Cumulative growth rate for eight consecutive years beginning from 1990 have been estimated at a bare 1.01 percent for Assam compared to 34.73 at all India levels. In a scenario where with each progressing year most of the other States of the country are performing better the case for Assam is the reverse. Not only has this feature persisted over a period of time, but also the intensity or severity with which it exists having such significant percent of population living in poverty as high as 45 and 40 percent in rural Assam for 1993–94 and 1999--2000 respectively is a significant point of interrogation. In the year 1999–2000 when Assam recorded the widest difference with the national averages on rural, urban and combined percent of population living below poverty line—13, 18 and 10 percent, respectively—the State also became the ground for maximum insurgent activities (Hazarika, 2001).

Despite the existence of many questions amidst various forces and 'will to question' the Government of India, there are other brigades of development programmes, some of which have already been launched and some others listed for launching in the next few years. The initiatives are manifestations of a predominant perception of the North East as an underdeveloped location, which needs to be developed through various programmes like globalization currently received as the most effective strategy. Among various ways of understanding the economies of the North East an immensely important and perhaps the most frequently utilised form of application in finding solution to the economic problems is the view that the region's landlocked economy and mountainous terrain create substantial barriers to domestic and international commerce. It also subscribes to the view that lack of development of the economies of the region is due to the absence of modern financial commercial, corporate institutions or due to institutions that are poorly developed. This is further endorsed by an additional view that the location of the region between three giant economies, those of China, South East Asia and mainland India and its histori straddling of North South trade routes, gives the region a potentialit to be a market-based economy. On the basis of these assumption there are various kinds of proposals for strategies of developmen solely and partly meant for the North East. To name a few, SAFTA initiative in South Asia, which began in 2006, Look East Policy focusing on cross-border land trade with China, Myanmar, Nepa and Bangladesh and on strong relations with ASEAN and the Asi Pacific Economic Cooperation group (APEC), Asian Highwa Project, an international highway system which will link the nationa corridors of India, China, Myanmar, Thailand, and Bangladesh alon with the Trans-Asian Railway, part of the Asian Land Transpo Infrastructure Development (ALTID) scheme that involves thirt two nations, Stillwell Road Project that could link the North East v

Myanmar into southern China etc. These projects, in one way or the other, enhance the authoritarian nature of the Indian State in the North East ever more as they involve heavy finance and coordination with other international agencies.

Conclusion

A simple question that everybody in the North East should now ask is: Do people really like the kind of development witnessed so far about sixty years? The answer lies in a proper evaluation of the developmental programmes in the region, which are merely superimposition of prescriptions. There are other ways to development and not simply following the dotted prescription of the international agencies that mostly are vehicles of global capitalism. Forest resources, water resources and rich biodiversity endowed in the region may offer different ways to development and of course an alternative development frame rather than going pompous about big industrialization and much hyped trade routes and zones via the North East. If this is not accepted to be the common principle of experiencing any kind of development in the region the cost will have to be borne only by the peoples of the region. While various forces of authoritarian presence of the State are fought back in recent years it is high time for the peoples of North East to stand together and raise questions about the connection between State atrocities and development programmes currently in operation.

REFERENCES

Bibhu Prasad Routray. 2001. *Neglected Northeast: Whose responsibility?* New Delhi: Institute for Peace and Conflict Studies.

DONER and World Bank. 2005. *Workshop Report*, November 10–11, Guwahati.

Hazarika, Sanjay. 2001. *The North East and the New Millennium. Dialogue and Conflict Resolution in North East.*

<http://ipcs.org/North_east_articles2.jspa?ction=showView&kValue=1154&status=article&mod=a

NER Databank. North Eastern Development Finance Corporation.

6

Understanding the Politics of Underdevelopment in the North East India

Roji Rajkumari

North East India (excluding Sikkim), covering a total area of 2,55,037 sq. km. inhabited by a population of 3.1 crore, belonging to a number of religious, linguistic and ethnic groups, remains a backward region. Although the region is full of unexplored potentialities, it continues to provide an interesting case of backwardness within backwardness. Development continues to be an issue relegated to the margins in the context of India's overriding obsession with 'security' or rather 'insecurity' as the region is always viewed from the stereotyped lens of a strategic buffer zone. Pervasive mainstream mindset examines the issue from the region's difference with the rest of the country compounded by the fact of sharing over 4,000 km. of India's International border with China, Myanmar, Bangladesh and Bhutan.

The energy-rich North East has substantial oil, natural gas, and coal exploited ever since British colonial period with substantial area left unexplored on account of numerous problems. Numerous rivers in the region flow enormous amount of water, and could generate far more electricity than they do. However, harnessing them raises environmental issues as well as political ones with international ramifications. The area also has abundant forest resources. It is nonetheless, India's most economically backward region. With virtually no substantial industrial infrastructure, industrialization is yet to dawn in the region even in an otherwise technologically driven world of the twenty-first century. On the contrary, deindustrialization seems to be the trend with ever-increasing dependency on mainland India.

In the post-British period, emerging political currents have created the base of ever increasing fissiparous tendencies worked and insinuated to the advantage of certain elitist interests. It is a bizarre sight to comprehend the region, which reported more than 200 schedule tribes apart from a number of ethnic groups not 'recognized' as such. In the midst of these dissipating tendencies, economic backwardness plays a crucial role in aggravating and exacerbating the fragile social fabric of the region. Economic backwardness is explicitly manifested in every aspect of people's lives. The deep sense of perpetual grudge in the minds of the people, antagonism and mistrust among various sections of the society, lack of a sense of belongingness and continuing communication gap aggravate the situation. The diversity of ethnicity and cultures, and the different levels of social, educational and political development in different areas of this region have all the more made the situation complex. In a way such differences have prevented the tribal people from coming upto the level of the people in the plains alongside the lopsided development policies undertaken so far.

Underdevelopment of North East : Dimensions and Factors

The present state of underdevelopment in the North East, which is due to the interplay of various factors—historical, geographical, psychological, social, political, cultural, economic, etc.—are interlinked with one another. Heightened proliferation of insurgency lends testimony to the fact that there is a historically inherent flawed integrationist policy along with a deep rooted sense of deprivation, frustration and feeling of 'otherness' among the people of the region. This feeling of otherness is reflected on a flawed process of nation making based on mistaken premises by the Indian ruling class. Assumptions devoid of historical realities, rooted in the legacy of the British Raj to which the Indian State stepped in to become the successor State have failed to evolve a sense of nationalist current found in mainland India. A persisting sense of 'otherness' remains intrinsic on both sides. Psychologically inherent sense of 'otherness' is intense due to the cultural, historical and racial specificities of the region. Difference seems to be pronounced on both sides as mutual exclusivity compounded by marginalization of the region on grounds of perceived difference remains grounded. The fact that this region represents the latest constituent of the Indian Union conditions the very psyche of its ruling elite. Subsequent responses of the Indian State to the various movements (based on the aspirations of the people) indicate lack of sincere understanding of the issues involved. Contemporary political history of the region in the post-British era shows that Indian ruling elite has willfully ignored or marginalized the aspirations of the people in the region. Pervasive nature of the marginalization, endemic in the ruling elite's psyche, forms the matrix of the polarized positions evident in the body politic of contemporary India *vis-à-vis* the conflict situation in the region.

A historical perspective to the problem would be crucial for an understanding. Widespread practice of blaming the Centre by the people for most of the problems and growing sense of alienation in post-British India also has its roots in the perpetuation of colonial policies. Persistence of colonial legacies in the policies has their impact on the people as they continue to think in terms of differences and conflict situations between the core and the fringe. As a result India has failed to evolve a post-colonial North East policy reflecting aspirations of the people in the region.

Independent India inherited a territorial boundary fixed by the exigencies of an alien imperialistic administration. Two crises were inevitable. First, ethnic groups were divided by the new 'boundary'. Secondly, modernizing elements were imposed. As a result of diffusionist modernizing wave there emerged a cultural crisis as the new mode had no practical relation with the prevailing mode of production of the 'native' people. The British policy was no doubt one of the least interference with regard to the hill people. However, the indigenous system of the region could not escape the impact of administration and its corollaries, the church, western education, legal institutions and most importantly, economic insinuations, which significantly destroyed their internal system of authority and exposed them to a different trajectory of power and politics. In practice colonial interest found direct intervention problematic and sought gradual transformation in the socio-economic arrangement of the societies inhabiting these regions. And as such, they maintained a sort of a distanced indifference from the tribal population.

Attempts seem to have been made although indirectly, during the colonial rule, to delineate the ethnic boundaries in terms of the 'Inner Line Regulation' and 'Scheduled District Act, 1874'. It is well known that segregation was the initial British policy for the frontiers. The inner line Act prohibited any subject living outside the area from living or moving therein. The Scheduled District Act was passed to set up an all-India pattern for the administration of the areas that might be declared 'backward tracts'. The Government of India Act of 1919 identified backward areas and sought to exclude them from the jurisdiction of constitutional reforms. The administration of the Government of India Act, 1935, divided the hills of Assam into two categories such as 'excluded' and 'partially excluded areas' (Bhattacharya, 1995, 19–20).

Although the mainstream leaders in the post-British period regarded this policy of 'excluded and partially excluded areas' as imperialistic and as such a part of the British Government's policy to alienate the tribes from the general population of the mainland by playing on their social and cultural distinctions, they themselves could not completely disregard the difference existing between the tribes and wider society after coming to power. The net result was that the colonial concept of exclusion found expression in the Constitution of India in the form of the Schedules V and VI, wherein certain mechanisms were evolved for administration of the tribal areas, which have not yet succeeded in accommodating the demands and aspirations of the people of this region.

Another historical explanation behind the failure of Indian democracy to accommodate the region may be rooted in the very nature and context under which India became independent. Indian polity is oriented towards the basis of a strong State rather than giving a space to smaller communities (Arora and Verney, 1995).

The strong centrist nature of the Constitution in whose making the voice of the region was wholly ignored is perceived as a threat to their own socio-political and historical specificities. Henceforth, the crux of the argument lies in the fact that very notion of a strong centrist State is incompatible with the multicultural diversities of the region (Chatterjee, 1997). Doctrines of 'equality of States and equality of nationalities' which are the backdrop of the federal polity elsewhere are yet to be translated into constitutional terms and doctrines, which should not, as such be contradictory to the political aspiration of the North-Eastern region for historical reasons. People in the region are yet to experience the sharing of political authority at the Centre on the basis of the universally accepted federal doctrines.

Flawed Legacies : Lacunae and Lapses

The present state of affairs in the country whose malaise is well known, as Sanajaoba has observed, is attributed to failures among many others on two broad fronts (1985, 249). First flaw pertains to the flaws in and the failure of the political institutions including the Constitutional relationship of the Federation with the constituent States, the concomitant misrule and mismanagement. And second flaw revolves around the pervading frustration and wilderness among the masses, who unfortunately, are rendered as scapegoats of all the failures in the governmental process. The problem of underdevelopment in the North East has not escaped these facts. The centralised system of planning seems to have constantly ignored local conditions and the Planning Commission has always shown an inclination for a uniform dull grey pattern of development all over the North East.

Every movement or voice of discontent is answered by the allocation of larger funds without in any way altering the pattern only led to the quicker siphoning off of funds back to the heartland with a few benefit accruing to those in power through the usual corrupt forces of the licence-permit Raj (Saigal, 1992, 215). This has led to increasingly corrupt regimes in States like Manipur,

Nagaland, Meghalaya and Assam. These regimes are widely regarded as representatives of the Central power to keep the people in a state of underdevelopment and many movements in the recent past have been directed against the so-called corruption and maladministration of the local governments and the indifference of the Central Government. The result leads to spiralling frustrations at the lack of progress in the region compared to the declared 'progress' in the rest of the country. The method adopted by the Planning Commission to assess regional developments in terms of per capita outlays, per capita incomes, etc. has tended to blur the systematic constraints such as notably hilly areas, long transport lead and lack of intense desire for a consumerist society which would otherwise have helped explain differential rates of growth (Saigal, 1992, 215). This has only succeeded in putting intolerable pressure on national resources and led to competitive politics without consideration of systemic absorptive capacities thereby making wastage and corruption the norm in the North East.

Outstanding aspects of the problem can also be traced to intrinsic defects of electoral politics. Frequent inter-party defections, existence of a larger number of national and regional political parties, temporary mergers and coalitions witnessed elsewhere also contribute significantly to the region's political instability (Ghosh, 2001). Abrupt transition from erstwhile self-governing social structures to a centralised administration with virtually little say in the affairs of the State leaves much room for discontentment. Ensuing civil strife and social unrest bog down the entire region with politics centered on petty ethnic issues of development. There is a need to rethink the idea of India becoming a more federal and democratic state with the growth of regional parties and subsequently, coalition politics and to analyse the perception of the smaller States of the region that have only one or two representatives (members) in the parliament.

The North East also suffers from a deep sense of economic neglect. A close examination of the turbulence and frustrations evident in the North East would indicate a number of underlying economic factors. However, it would not be an exaggeration to go with the widespread feeling that pattern of development to which the North East in general has been exposed, is purely extractive, exploitative and is 'colonial' with little reinvestment and profits. Therefore, a more closed economy might yield better dividends while protecting identity at the same time. It has been argued in mainstream Indian economic discourse that the North East has not been neglected in terms of money. On the contrary, considerable sums have been poured into the region without proper accountability which in turn consolidates the politicians-bureaucrats-contractors nexus and interest (perceived widely as representatives of the system) at the expense of ever increasing dependency as productive forces are thwarted and obliterated. The issue of 'easy money' is widely heard all over the North East. This money comes from the Centre and has in good part been drained away by a small oligarchy or elïte, the consumption of which contrasts sharply with the general level of development.

Education has been accepted as a crucial factor in bringing development. The literacy rate in most of the North-Eastern States is above the national average. For example, Mizoram is second to Kerala in terms of literacy rate. However, the figures have not been translated to economics development so far. The educated youth in turn finds fruitful employment difficult. The educational system acutely lacks value orientation. Quality educational infrastructure is still rare and remains a distant dream in the region. There is hardly any substantial infrastructure for high quality technical and vocational education. In such a situation, those who can afford send their wards for education outside the region. However, this causes an enormous resource drain on the region. On the other hand, there hardly exists any prospect for those in vocational streams or entrepreneurship as industrial infrastructure virtually does not exist. The curriculum does not reflect the requirements of the region. It is merely a reproduction of the curriculum set by those at the centre.

Moreover, lots of consumerist culture in the North East has severely affected the largely self-sufficient tribal economy. Absence of the secondary sector leads to massive dependency on the heavily tertiarised public sector for employment. There is an urgent need to formulate policy measures, which aim at creating productive bases. Such a formula calls for a proper understanding of policy makers to work out effective measures rooted in the ground realities.

Low investment in productive jobs is an important issue. The North East is treated only as a supplier of raw material of petroleum, tea and coal to the rest of India and consumer of finished products (Baruah, 1999). It is ironical that a State such as Orissa, which is considered backward, has comparatively more industry than the entire North East put together. Without justifying the happenings, an important thing to be focused is the failure of the economic forces to create jobs in the region to the level of educational attainment of the people.

Spatial Configurations : Isolation at Work

Geographical setting of the North East along with interplay of social, cultural and economic forces between the different ethnic groups also provides much of the explanation of the recurrent period of crisis and turmoil to which the region is subjected. Link with the central heartland tends to be weak geographically, economically, socially or culturally. The region is connected with the rest of the country by the narrow 'Siliguri Neck' of less than 20 km. wide. Transport link is not tedious and weak but also 'unnatural'. The 'normal' flow of goods is through Bangladesh for South Assam, Mizoram and Tripura and Myanmar for large parts of Manipur and Mizoram. Long and arduous transport link only means higher costs for consumables but also corresponding low prices for the products of the region thereby leading to very adverse terms of trade. The picture of development in other parts of the country adds to the existing frustrations and feeling of being left behind.

Apart from a distinct geo-political environment, the region also has a peculiar demographic formation. The indigenous population has been exposed naked to colossal migration waves from outside. It brings us to the related issue of immigrants. The focus has been on the Bangladeshi immigrants. But studies indicate that at least half, if not two-third, of more than three million immigrants are from the Hindi heartland, mostly from Bihar and Uttar Pradesh, and West Bengal (Fernandes, 2004). They come as cheap labourers. Even where the locals are ready to work, the employers prefer immigrants because they work for lower wages. Under such circumstances, the native people of the region start developing a feeling of revulsion towards the outsiders. In this context, the people have developed rigid and exclusive identities within the region. People living in this part of the country see any trespassing on their land with suspicion. The people also resent the control of their economy by outsiders.

In addition to the general problems, the North Eastern region has certain specific problems as well. These problems of the region have grown out of proportion, partly due to the congenital transmission and partly due to prolonged negligence and disinformation by the Central Government (Sanajaoba, 1985, 256). Abrupt imposition of mainstream Indian politics with all the vices and little virtues affects the simple but highly unstratified body politics of the small ethnics and tribes. The social frictions which were absent in economic terms are verbal, loud and becoming more articulate. Moreover, the socio-cultural stability of the regional ethnics has been profoundly affected by the unchecked influx of outsiders for the last five odd decades continuously. Relatively stable conditions are affected by the trend of displacing indigenous ethnics by alien migration, the changing value patterns, the central administration oriented in the line of internal colonialism, the prolonged negligence and emerging socio-political organization and the newly invented social stratification among others. The social and political corollary of such disturbing trends, which adversely affect the political culture of the ethnics, had been outside the perception, experience and vision of mainstream politics. If the Central Government treats the issues of the region as inevitable, it should be blamed for all the malaise that grips the region (Sanajaoba, 1985, 259).

The region also experiences widespread insurgency. Although each conflict has its own roots and history, the issues they raise include language and ethnicity, tribal rivalry, migration, control over local resources, access to water and feeling of exploitation and alienation from the Indian State, and most importantly, independence and autonomy (Hazarika, 1995). These movements have been given different causes and meaning from time to time by the Indian Government, sometimes to the extent of stating involvement of foreign powers going beyond domestic discontent (Verghese, 1996, 335). Aspirations of the people in the region from their 'merger' with the successor Indian State, after British departure from the sub-continent are yet to be realised meaningfully. Contemporary developments bear solid testimony to the fact that grievances of the people in the region are genuine. Fear for loss of ethnic identities and agony of poverty in their affluent land are the two-most dreaded devils in the region (Sanajaoba, 1985, 252). There is compelling need to look at the situation beyond the military view.

Economic Planning : Need for Rethinking

The economic development plans have also been such as to allow the ruling elite to scheme off a substantial part of the expenditure. Within the framework of the Constitution, it tried many experiments —District Councils, a State within a State, a State with power reserved for the Governor to tackle law and order, a regional council (NEC) for development and so on. However, in the absence of innovative economic and social policies, all these soon ran out of steam. Besides, attitude and processes remained unaltered thereby generating tremendous pressure on structures (Saigal, 1992, 228-229).

There is a general impression that the central polity lacks the required will and ability to stand up to sustain agitation or insurgency. The policy pattern followed in the North East is inadequate and partial and in no case was the impression given that the Central Government was committed to genuine development. On the other hand, by focusing on elite requirements and desires it brought into sharp focus the low availability of government and other public sector employments (Saigal, 1992, 230). It leads to scuttle among different groups for this scarce commodity and intensify existing ethnic differences and tension. The federal system may be successful in containing insurgency and other law and order situations to a certain extent but has failed miserably in the economic, cultural and social spheres. Militarization and use of brutal force will not erase insurgency as the basic roots of such a problem come out to be one or more elements or combination of many which are related to socio-cultural, economic, political and most importantly historical issues. Therefore, counter insurgency strategies will have to be a combination of efforts. There is a crucial need to evaluate whether the problem of insurgency is to be seen within the parameters of economic backwardness, unemployment, corruption or certain other enduring normative elements associated with historical aspirations.

Conclusion

New Delhi persists in treating insurgency as a law and order issue or as instigations by foreign hands with communal tones by focusing on Bangladeshi immigrants while ignoring those from the Hindi heartland. Another reaction is announcing of crores of rupees to be poured into the region without any investment plans for generating productive forces. This policy matrix has only exacerbates corruption as politicians, bureaucrats and contractors appropriate what comes into the region. With no follow up in the form of investment plans, nothing comes up. No production based industries that can tap resources are created. It leads to further disillusionment. Neither the communal bias nor the exclusive focus on law and order can take the region out of this trap. Both the Centre and the local leaders have to concentrate on regenerating the economy and creating low investment productive bases. On the contrary, the Centre plans to turn the region into a power-house to supply power to the rest of India by building more than 40 massive dams during the next decade. The result will be colossal loss of means of livelihood to the already marginalized millions of North East region instead of development. For such dams will destroy forest and environment that hither provides suitable means of livelihood to the tribal people. Such policy will certainly result into further marginalization.

In the context of policy making, the people of the region should find expression at every decision making point of the Central

Government and State Governments. The Union Government is yet to formulate and implement a sound and equitable long-term region policy and address the aspiration of the region in durable terms. Failure in this front leads to nearly 3,000 annual ethnic explosions (Sanajaoba, 1985, 253). The lags are sufficiently large and pressing to warrant a clear focal point in Delhi to project the needs and push the implementation of programmes in the region (Sachdeva, 2000). With India moving rapidly into a new era of globalization propelled economic reforms, there is serious need to interrogate the proposed integration and opening up to the economies of East and South East Asia. In other words, India needs to look beyond a century old vision based on the frontier concept held by the British colonial apparatus.

REFERENCES

Arora, *Balveer and Verney, Douglas*. eds. 1995. *Multiple Identities in a Single State: Indian Federalism in Comparative Perspective*. New Delhi: Konark.

Baruah, Sanjiv. 1999. *India against Itself: Assam and Politics of Nationality*. Philadelphia: University of Philadelphia Press.

Bhattacharji, J.B. 1995. Northeast India perspectives in history. In *Northeast and the Indian State: Paradoxes of a Periphery*, ed. P.S. Data. New Delhi: Vikas.

Chandhoke, Neera. 2006. A State of One's Own: Secessionism and Federalism in India. <www.crisis states.com,September>

Chatterjee, Partha. 1997. *State and Politics in India*. Delhi: Oxford University Press.

Fernandes, Walter. 2004. Assam is Burning. *The Hindu*. November 25.

Ghosh, Subir. 2001. *Frontier Travails: Northeast: The Politics of a Mess*. New Delhi: Macmillan.

Hazarika, Sanjoy. 1995. *Stranger of the Mist: Tales of War and Peace from India's Northeast*. New Delhi: Penguin Books.

Sachdeva, Gulshan. 2000. *Economy of the Northeast: Policy, Present Condition and Future Possibilities*. Delhi: Konark Publisher.

Saigal, K. 1992. Federal democracy and pluralism in the Northeast. In *Federalism in India: Origin and Development*, ed. Nirmal Mukarji and Balveer Arora. New Delhi: Vikas.

Sanajaoba, Naorem. ed. 1985. *Basic Issues on Center-State Relation*. New Delhi: Omsons.

Verghese, B.G. 1996. *India's Northeast Resurgent*. Delhi: Konark.

7

Economics of Underdevelopment : Reading the North East

Thingnam Kishan Singh

History of economic development is as old as the history of human society. As human existence progressed, development has remained a central concern. What necessarily constitute development or the parameters of development or how do we understand development contextually has perhaps been one of the most enigmatic issues in the domain of human and social sciences. As human beings organize themselves into communities and societies with characteristic features of production based on surrounding environment development inherently deposits the centrality of production that sustains human life. Modes of production, therefore, are necessarily correlates of modes of development. Change in the mode of production inevitably conditions the nature, scope and impact of development.

Development debates *vis-à-vis* the North East generates polarized positions. What has perhaps been postulated as complex as far as development debates of the region are concerned conspicuously underlines the marginality of the debates themselves. Purported complexity of the issue has been an overarching articulation amongst the policy makers in New Delhi. At the outset, it is important to note that conceptualizing the region as a compact entity is not without problems. However, location, history and culture have contributed to the region's characteristic distinctiveness where difference with mainland India is at once immediate and striking. Given the historical evolution of the region as an entity in terms of the geographic, political, economic and social space in occupies, modern understanding of the region apart from its historico-cultural configurations is intertwined with colonial geography and historiography. Even since British colonialism took roots in Bengal in mid eighteenth century a systematic attempt to explore the 'frontier' of Bengal remained an obsessive concern for the colonial administration. As it was used to be known during those days as the 'frontier region' of Bengal was loaded with things important for the 'empire' (Mackenzie, 1884) (Reid, 1942).

Rivalry among European colonial powers over colonial expansion led the British to concentrate initially on mainland India with the coastal regions perceived as strategic both in terms of economics and politics in line with its maritime framework. However as colonial expansion proliferated across the length and breadth of the Asian continent, increased and intensified competition between the English, French and Dutch in South East Asia had repercussions in this otherwise 'wild' region marked by 'inhospitable terrain' with mountains, deep gorges, ridges, hills and villages. Consolidation of colonial domination in mainland India along with expansionist threat posed by the French in Indo-China led the British transform their exploration of the region to the perpetration of the colonizing process. As the nineteenth century progressed the British colonial authorities in mainland India grew extremely conscious of the strategic importance of the region (Mackenzie, 1884) (Reid, 1942). Since the outbreak of the Anglo-Burmese war in 1626, the British Raj made a conspicuous attempt to expand eastwards. Ahom (Assam) was the first casualty as the British soon overran the fertile Brahmaputra and Surma valleys. With the signing of the Treaty of Yandaboo in 1826, Manipur also became an important entity for the British. This moment of colonial impact triggered a series of events leading to drastic changes in the region. It was crucial for the colonial regime to incorporate this region in its colonial domain so as to open the vital transit route to Burma and Thailand, which eventually became part of the British overseas empire.

With the extension of colonial domination to the kingdom of Ahom in the first-half of the nineteenth century, a gradual process went hand in hand with the subjugation of the 'wild tribes' in the mountains of the region such as Lotha Nagas in 1875, Angami Nagas in Angami Hill in 1878–80, Ao Nagas in 1889 and Lushais in 1871. Colonial geography replete with colonial cartography systematically organized administration of the region to facilitate economic exploitation. Territoriality of the empire inscribed in colonial geography was thereby responsible for constructing the dynamics of a political economy marked in terms of an idea articulated as a region. Transformation of the whole region into a frontier apparatus of the colonial possession in India led the British to extend their influence and domination over the entire region. Inscribed in this colonial geography was the visualization of the region as a frontier area for their prized possession known as 'the jewel in the British Crown' (Reid, 1942).

It soon dawned upon the colonial apparatus to explore the necessity of exploiting the immense resources of the region. Plundering of the resources in the Brahmaputra and Surma valleys got underway as the British realized that it could add substantially to their coffer. In the process, Ahom witnessed a kind of colonial domination more intense than other parts of the region (Guha, 1991). Even since the first-half of the nineteenth century the Lakhimpur district in Assam witnessed frantic settlement programmes undertaken by the British to exploit the tea potential. Tea bushes were discovered in the area, which was then sparsely populated. As a cash crop tea was responsible for systematic and organized expansion of the plantation economy whereby immense British capital could find profitable investment. Vast tracts of uninhabited land throughout the Brahmaputra valley were soon reclaimed and acquired through meticulously crafted land reform policies meant to subserve the interests of the colonial power. 'Tea mania' led to unchecked expansion of tea plantations with easy access to land dexterously dispossessed from the indigenous population. Speculative aspect associated with this mad rush for land for tea plantations undermined and affected productivity in the region. Massive surplus thus extracted from the plantations were exported outside the region to different parts of mainland India and elsewhere in Europe (Ahmad and Biswas, 2004). Colonial economies thereby created foundations for underdevelopment in the region.

The neighbouring areas were marked as the North East Frontier Agency (NEFA) with substantially lesser interest taken by the colonial authority (Elwin, 1964). Leaving the extremely difficult upper ridges of the mountains the adjoining areas known as NEFA witnessed the introduction of a monetized economy hitherto unseen before. Traditional trade carried on under an equitable exchange system suffered immediate rupture as a system of unequal exchange based on cheap manufactured consumer items flooded the hill areas. A penchant for these otherwise frivolous items grew alongside the rapid strides made by cultural imperialism as the tribes become increasingly dependent on imported commodities. In the meantime, a merchant class of Marwaris and Bengalis from mainland India subserving the interest of the colonizers were introduced to facilitate monetization of tribal economy in the region. This merchant class introduced opium and carried on a brisk trade with far reaching consequences for the people in the hills as they were made to trade natural products like timbers and rubber for opium. Dependency thus grew amongst the tribes for manufactured commodities from outside and along with consumption of opium destroyed the traditional economy and depleted traditional resource base (Ahmad and Biswas, 2004, 52). Integration of the economy of the region into the world capitalist system through the mechanics of colonial imperialism led to its transformation into a peripheral underdeveloped economy.

Even as India started to embark on the process of industrialization during the latter half of the colonial period due to concomitant and corollary efforts of the British and the emerging Indian capitalist class transformed from its feudal base—there was a conspicuous absence of similar developments in the North East. British colonial policy of visualizing the region as a frontier periphery largely contributed to the undermining of any productive based effort in the North East. Colonial interests in the region were served incredibly well by the merchant Marwari class who exported the surplus of the region. The smooth running colonial economy through the facilitating role played by these Marwaris who extracted and exported surplus could not afford jeopardization as far as any substantive attempts to develop the industrial potential of the region was concerned. Mention may be made of the surreptitious role played by the Marwaris who in their extraction and export dynamics created artificial scarcity of rice in Manipur in 1939—the year in which production level was highest—which is turn led to the Second Women's War (*Nupilan*) (Singh, 2006). Surplus export had touched new levels whereby artificial scarcity was created. Close survey of the economic undercurrents in the region since the British colonial intrusion reveals the steady emergence of the region's economy as a peripheral capitalist economy *vis-à-vis* its integration into the world capitalist system during the colonial period. It was essentially during this period that the economy of the North East underwent a transition from its various pre-capitalist social formations to a quasi-capitalist organization of productive forces. It can, therefore, be argued that such a transition was not a result of strictly chronological features.

The region's historical development and progress of its mode of production have not yet had the necessary material base nor the time required to transform itself successfully into a capitalist system. What happened was the thwarting of the regions potential to transform its productive forces in order to evolve a truly meaningful development framework of its own. Instead, it witnessed the introduction of an economics that benefited the colonizers and the mercantile class of outsiders comprising of Marwaris and Bengalis. Traditional resource base and economic modes in the tribal areas were ravaged with their integration into the colonial economy. Destruction of the fragile tribal economy and subsequent emergence of a new base for creation of a modern economy based on colonial exploitation characterized the hill areas in the North East. In the area marked as NEFA traditional economic arrangements like the *posha* through which the tribal people levied taxes on material goods of the plains' people rapidly eroded as a result of the introduction of a monetized economy. The tribal in the hills were soon deprived of the right to levy and collect taxes. Market relations introduced by the colonial authorities along with the steady implantation of a trading class comprising of Marwaris and Bengalis ruptured the traditional network in the region. Distribution, circulation and exchange of products between the hills and plains differed immensely (Mishra, 1983). These transformations were instrumental in converting the region into a peripheral economy replete with the system of uneven trade devised by the British policy of opening up of local barriers.

Obliteration of the traditional barriers that gradually led to the influx of immigrations introduced as a part of the colonial class to facilitate the transformations through an intermediary class discouraged local entrepreneurship (Ahmad and Biswas, 2004, 53). Enjoying patronage of the colonial authorities and superior financial and material resources and networking relations these absentee interests soon controlled the entire economic landscape of the region. Meticulously crafted colonial policies surreptitiously made rapid inroads to increase dependency of the tribes on unequal trade with the emerging commercial class to subserve the interest of exerting pressure to extract concessions for colonial expansion (Sikadar, 1982). End of British colonial domination in the subcontinent in the post-Second World War era did not produce any path breaking changes as the region continued to reel under similar conditions in the face of an external market, now in the form of the mainland India market and its dominant capitalist class (Mohapatra, 2002, 24). Integration of the North East with the successor Indian state in the post-British period—starting with the 'merger' of Assam, Manipur and Tripura in 1949—failed to produce the dynamics of a successful transformation in terms of economic development. Mere semblance of a socialist pattern through indicative centralized planning over the last five and a half decades miserably failed to generate development and transform the region's backward and underdeveloped economy.

Reasons are not far to seek. As decolonization dawned in the post-war years it remained imperative for the newly decolonized to choose between two modes of development—the capitalist and the socialist. India's capitalist logic of development—evident in its indicative planning and mixed economy with massive deficit financing of the private sector by the state—concentrated and confined development in certain industrialized metropolitan pockets. It has been widely argued that the Indian state perpetuated the legacy of colonial economy with the capitalist mode of production in the post-British period as a result of the alliance and collaboration between western metropolitan capital and the indigenous capitalist class (Patnaik, 1999). The Indian indigenous capitalist class located in the industrialized locations in the mainland had already made significant inroads in the colonial economy during British rule. It is pertinent to note that British departure did not entail a similar picture as far as Western capital was concerned. Indian economy was still under impact at capital investment from Britain and other Western capitalist powers. Western capital still had immensely high stakes in India. Significantly, it was the successor Indian state that became instrumental in propelling capitalist development by evolving complex strategies for achieving growth of capitalism in India (Bhambhri, 1988, 6). Mobilization of resources on a large scale for investment and the resultant high capital-output ratio reveal the logic of capitalist contradictions. Capital driven development was concentrated in certain specific locations in mainland India while the North East as a region continued to suffer from a perpetuation of the logic of underdevelopment. Development issues *per se* in the North East continue to reflect the structural inequalities of patterns of economic relationship with both the global capitalist system and the newly emerged Indian capitalist system. State intervention in the post-integration period has been directed from the top. It is an explicit superstructural intervention without a firm base transformation. The superstructural excrescence resulting in stark disparities in terms of economic development can be seen clearly. Development debates on the North East in mainstream Indian intellectual discourse, which ironically exudes a journalistic and bureaucratic slant reflective of the work done by journalists and bureaucrats posted in the region, have conspicuously ignored the inextricably intertwined relationship between modes of development and modes of production. In a characteristic mould, economic backwardness in the region has been seen and analysed through the lens of *a priori* integrationist model over zealously and fervently working towards the creation of a monolithic discourse. Insularity has been a characteristic hallmark of these analyses on the region's economic standing as the basic issue of production was conspicuously overlooked. Beyond the stereotyped premises of 'neglect' or massive allocation of funds through central assistance, serious study of the region's productive management is yet to be seen. The core issue of production engaging different aspects of production sectors and potential has largely remained on the margins of development debates. Economic development sans production issue can only generate haphazard policies which in turn will eventually render multi million economic packages and central assistance lopsided and meaningless. Underdevelopment of productive forces in the region is something that needs a detailed analysis to understand the failures behind the much hyped economic policies and packages designed in the power corridors of New Delhi.

Failure of economic policies and packages in the North East can be perceived strikingly clear from a cursory glance at some of the basic indices commonly used to measure the state of economic and social development. The per capita net State domestic product of the constituent States of the region indicates a dismal picture of disparity in comparison with the all India average. Massive underdevelopment formed the matrix of the fact that when India's gross domestic product was growing at an average over six percent in real terms, economies of the region were growing at a lesser rate during 1992–1999. NSDP of the seven States of the region has shown a marked decline over the years. Decline in the annual average growth rate of NSDP has been ascribed to the increase in dependent accumulation in the region (Ahmad and Biswas, 2004, 63).

Even as GDP is known to hide many things, a low NSDP is not considered congenial for improving the living conditions of the people. With a primarily agrarian based economy that still caters to subsistence production rooted in obsolete methods of farming, the unmechanised and uncommercial nature of agriculture in the region is a sector on which more than seventy percent of the population depends upon. Population engaged in agricultural activities for livelihood constitutes the majority showing intense disparity with the all India average. High dependence on an otherwise primitive agriculture based on subsistence level of production indicates the nature of the underdeveloped economy of the region. Increasing pressure on land as a result of population growth compounded by the quantum leap through massive influx of immigrants from Bihar, Bengal, Orissa, Rajasthan, Punjab, Bangladesh and Nepal and recent trends in the wake of globalization triggered liberalization transforming cultivation configurations from food crops to cash crops contribute to significant increase in dependency. Increasing trends of displacement of rural cultivators as a result of haphazard urbanization without concomitant creation of productive base have been widely noticed. Dependency is aggravated and exacerbated in the face of decreasing production levels in agriculture, which ironically constitute the mainstay of the region's economy. Dwindling productivity in the primary sector in turn increases dependency. Studies conducted on production in the primary sector have shown how a traditionally surplus exporting economy like Manipur—even during and after British colonial occupation—has been transformed into a net importer. Rice as the staple foodgrain cultivated in Manipur always reached surplus production levels with subsequent export of the surplus. However, ever since Manipur's 'integration' into India in the post-British period, alarming decrease in production levels has been noticed with its subsequent transformation into a rice importing State where twenty percent of its requirement is imported from mainland India (Chongtham, 2005).

Despite the polemics surrounding the use of per capita income as an important indicator in assessing poverty and economic backwardness—its limitations notwithstanding—it is, however, generally true that people in areas with low per capita do not live well. The per capita picture in India reveals striking disparities with the States in the North East displaying pathetic levels. Generally speaking, a cursory glance at basic statistics speaks volumes for increasing dependency of the region. Per capita income in Assam in 1950–51 was found to be 4 percent above the all India average marking it as one of the most prosperous State in the entire eastern belt. However, decades of centralized planning have led to a drastic fall showing it at 41 percent below the all India average in 1998-99 at current prices and 45 percent below the all India average at 1980-81 prices. While per capita income at 1980 prices grew by 40 percent for all India during 1980-1990, Assam's grew by only 20 percent. When all India growth rate of per capita was 37 percent during 1980–1998 in Assam it grew by just 10 percent (Government of India, 2003). Similar picture avails in the other States of the region.

Majority of the population in the region still live below the poverty line. Mass poverty as an acute problem is complementary to income inequalities. Estimate based on the study by a Planning Commission Group headed by Lakdawala gives a figure of 36.5 percent of the total population of the region living below the poverty line. Using an important index of human poverty Index (HPI)—a composite index frame on the basis of deprivation in economic, educational and health sectors all equally and proportionally given due weightage, the North East region represents a dismal picture in both the urban and rural areas. Drawing the comparable data of 1981 and 1991, there is an insignificant or marginal improvement in Assam, Arunachal Pradesh and Mizoram while the other States have significantly declined rapidly (Government of India, 2002).

Massive underdevelopment in the region is compounded by the increasing trend of deindustrialization. In States like Manipur all the districts fall under the category of 'No Industry District' by all India standards (Hanjabam, 2006). Virtual absence of industrial infrastructure, compelling the region to depend only on the primary sector of production, even after five decades of centralized Indian economic planning conspicuously narrates the saga of underdevelopment. Starting with the Second Five Year Plan more widely known as the Mahalanobis Plan for intense industrialization in India, economic planning has never relegated the importance of creating industrial base to modernize the economy. However, in the North East increasing deindustrialization presents a different story altogether (Chongtham, 2005). Sheer absence of industrial development and production units in the secondary sector indicate the persisting peripheral status of the region's economy. Increasing dependency marks the economy of the region with heavy imports of consumer and non-consumer commodities from the mainland which in turn virtually make it impossible for the region to carve a path for economic self-reliance. It has been estimated by the Shukla Commission that over Rs. 2,500 crore worth of consumer items are imported from mainland India annually. Glaring instances of dependency are provided when Assam, the most 'industrialised' State in the region, imports seventy-four percent of its milk product, ninety-eight percent of its meat requirements and 94 percent of its eggs from outside the region. In spite of heavy rainfall and natural abundance of water resources, the State imports half of its fish requirements from outside (Ahmad and Biswas, 2004, 56). Manipur, traditionally self-sufficient in fish production with a large variety (Brown, 1873) today imports more than three-fourths of its requirement from outside, the bulk coming all the way from Andhra Pradesh. It is indeed a bizarre sight to see that a predominantly rural Manipur has to import nearly all its eggs and poultry requirements from a distant Andhra Pradesh. The fledging poultry sector in the State catering mostly to meat production in also filled with incongruous terms of trade where one to seven days old chick are flown in by air, supplemented by all the feed items right from the initial to the consumption stage coming from States like Orissa and West Bengal. This anomaly renders it impossible for local entrepreneurs to successfully develop a self-reliant poultry. It has also been noticed that the low credit-deposit ratio in the region leads to the transfer of rupees 5,000 crore annually to other regions for investment by the banking sector. Lopsided and haphazard development policies rafted in the ivory towers of New Delhi detach the issue of production from development. Thwarting of local productive forces and failure to develop productive base tend to increase dependency on the mainland. Transformation of the region into a market for manufactured goods from mainland India went hand and hand with increasing underdevelopment in spite of sheer abundance of prime resources like oil, timber, tea, gas, limestone and other forest resources (Ahmad and Biswas, 2004, 55).

The predominant feature of underdevelopment, evident in the conspicuous absence of industrial and production base in the region, has been conveniently ignored in mainstream development debates. Uneven and unequal regimes of exchanges seen so far ironically subserve the interests of capital elsewhere representing the legacy of underdevelopment of a region considered a frontier in so many ways. Even since the British colonial period, the oil-bearing area in Assam has seen a meticulous exploitation of this precious resource (Edney, 1997). The British knew of the existence of oil in the hills of upper Assam as early as 1825 and identified the oil bearing belt which runs from extreme north east of Assam to the eastern border of the Brahmaputra and Surma Valley. Drilling of oil commenced in 1866 and 1865. The British company McKellop Steward Company drilled many shallows wells near Makum during 1866–68. Soon it led to the discovery of Digboi oilfields, the biggest oilfield in the subcontinent. Annually it raises about four lakh tonnes of crude oil. Situated 30 km. away from Digboi is the Naharkatiya oilfield discovered by Assam Oil Company, an Indian Government undertaking, in 1955. Naharkatiya is the second largest oilfield in India with oil reserves estimated to be about five million tonnes. Bappapung, Hansanpung, Hugirijan and Moran, located about 40 km. south west of Digboi, are also important oilfields. Mention may be made of the fact that quest for oil formed an important part of the colonial enterprise in the region.

Post-British oil industry never witnessed any radical departure, as Indian State monopoly was the only significant replacement that never really percolated to the basic foundation of the region's underdeveloped economy. It remains a sector existing to meet demands elsewhere on unequal terms of exchange. Set up in 1901, the Digboi refinery has not grown in terms of installed capacity even as it continues to be the main and the most important supplier of crude refined elsewhere. Its installed capacity is a mere 5.20

lakh tonnes. The Noonmati refinery near Guwahati under the Indian State owned Indian oil Corporation (IOC) with an installed capacity of just 8.8 lakh tonnes and the Bongaigon refinery set up in 1978 west of Guwahati with a capacity of 10 lakh tonnes stand in marked contrast with the Barauni refining in Bihar, a non-oil producing area, set up in collaboration with USSR. Production in the refinery began in 1964 with a refining capacity of 33 lakh tonnes, the rider being its total crude receipt from the oilfields in Assam.

The logic of ruthless plundering of the resources without proportionate returns to the people of the region can be substantiated by the sheer discriminatory rates of revenues between the Centre and State. A cursory glance at the revenue sharing rates speaks volumes for the nature of resource plundering. The royalty Assam got for its oil during the mid-eighties witnessed a marginal increase from Rs. 325 per tonne to its current rate at Rs. 632 per tonne in comparison to the centre's earning at an average dividend of Rs. 2,623 per tonne. It is equally distressing to learn that two-third of Assam's crude oil is transported outside the region for refining. In the midst of widespread demand for an optimum capacity refinery to refine its crude, Assam got a toy refinery at Numaligarh (Biswas, 2006, 235).

The tea industry in Assam also experienced similar patterns of extraction and exploitation. Even since discovery of tea by the British, a mad rush for plantation culture proliferated with tremendous growth of tea estates. Departure of the British eventually saw the exit of European companies with subsequent increase in the number of estates. Statistics reveal the rise of tea estates from 785 with an acreage of 155,674 hectares producing 15,03,70,000 kg. in 1951 to more than 848 tea estates with an acreage of 2,36,811 hectares producing 40,29,47,000 kg. in 1993 (Nag 2006, 130). Drain of profit in the tea industry is also colossal. Figures indicate that Assam produced an average of 250 million kg. annually during 1962–1999 at international price estimate of US$ 2.5. However, it is distressing to learn that the tea companies—almost all of them owned by outsiders—have invested almost nothing in the State from this staggering amount. These are ample evidence of the ways in which the resources of the North East are drained out and transformed into sources of profit in elsewhere whole simultancously unleashing a chain of exploitations networks shifting the development outputs from the region (Biswas, 2006, 235).

Forest resources—found in abundance in the region—have been similarly plundered and exploited ruthlessly. Assam, Arunachal Pradesh, Nagaland, Meghalaya and Manipur have suffered the consequences of mindless and ruthless logging and lumbering business catering commercial interests of outsiders. With the erosion of traditional indigenous modes of replenishment and aforestation in these traditionally forest based economies along with the displacement of tribal population as a result of forest laws enacted in the distant power corridors of New Delhi, the region has witnessed an upheaval in the face of the State exercising absolute monopoly of control. The nature of ruthless plundering and exploitation can be comprehended by the virtual absence of any substantial forest based industry in the region except for the odd miniscule saw mills and a few paper mills in Assam. Timber from the region feed the plywood industries in West Bengal, Delhi, Uttar Pradesh, Haryana, etc. The finished products find their way in the domestic and international markets. In typical peripheralization fashion, the region still performs the role of a supplier of primary products to feed the industries located elsewhere in different parts of mainland India. Paper, plywood and other forest based products like, paper gum, resin, lac and all rubber items have to be imported thereby displaying the classic *modus operandi* of transforming the region into a monocultural economy. Diversification, that can be brought about only with substantial industrialization, is nowhere on the horizon. The nature of development policies, therefore, has not all addressed the basic and fundamental issue of production in the region. If production and modes of production in the region remain confined to the monoculture level—where the region only acts as a supplier of primary products in these three sectors, oil, timber and tea—development discourse will remain lopsided and haphazard. Development, therefore, should necessarily address production and generation of productive base in the region. Only when the region emerges as a strong producing base rooted in using the resources available for the benefit of its people, development semantics will assume a holistic nature. It will enable the region to utilize its resources meaningfully and erase dependency.

REFERENCES

Ahmad, Raiful and Prasenjit Biswas. 2004. *Political Underdevelopment of Northeast India*. New Delhi. Akansha Publishing House.

Bhambhri, C.P. 1998. *Politics in India 1947–1987*. New Delhi: Vikas.

Biswas, Prasenjit. 2006. Development as 'complementary': A political economy critique of policies and practices of development. In *Challenges of Development in North-East India*, eds. David R. Syiemlieh, 208–58. New Delhi: Regency Publications.

Brown, R. 1873. *Statistical Account of the Native State of Manipur and the Hill Territory under its Rule*. Calcutta: Office of the Superintendent of Government Printing.

Chongtham, Priyoranjan. 2005. Manipur's Economy: Historical Roots and Structural Evolution. *Eastern Quarterly* 3, III. October–December: 149–59.

Edney, M.H. 1997. *Mapping an Empire: The Geographical Construction of British India 1765–1843*. Chicago: Chicago University Press.

Elwin, Verrier. 1964. *A Philosophy of NEFA*. Second reprint. Shillong: Advisor to the Governor of Assam.

Government of India. 2002. *Compiled National Human Development Report 2001*. New Delhi: Planning Commission.

Government of India. 2003. *Assam Development Report*, 2002. New Delhi: Planning Commission.

Guha, Amalendu. 1991. *Medieval and Early Colonial Assam: Society, Polity, and Economy*. Calcutta: Centre for Studies in Social Sciences.

Hanjabam, Isworchandra Sharma. 2006. India's Look East Policy and Manipur's Economy: A Critical Scrutiny. *Alternative Perspectives* 1, IV. July–September: 55–70.

Mackanzie, Alexander. 1884. *History of the Relatives of the Government with the Hill Tribes of the North-East Frontier of Bengal*. Calcutta: Superintendent of Printing.

Mohapatra, A.C. 2002. Development and underdevelopment in the Northeast region: Search for a paradigm. In. *Development Priorities in North*

East India, ed. Bimal J. Deb, 23–30. New Delhi: Concept Publishing Company.

Mishra, S.N. 1983. Arunachal's tribal economic formations and their dissolution. *Economic and Political Weekly*. Vol. XVIII, No. 43, October 22: 1837–1846.

Nag, Sajal. 2006. Land, migrants, hegemony: The politics of demography in Northeast India. In *Challenges of Development in North-East India*, eds. David R. Syiemlieh, 126–45. New Delhi: Regency Publications.

Patnaik, Prabhat. 1999. *Whatever Happened to Imperialism and Other Essays*. Delhi: Tulika.

Reid, Robert. 1942. *History of the Frontier Areas Bordering on Assam from 1883–1941*. Shillong: Government Publication.

Sikadar, S. 1982. Tribalism *vs.* Colonialism: British Capitalistic Intervention and Transformation of Primitive Economy of Arunachal Pradesh in nineteenth century. *Social Scientist* 10, 115. December: 15–31.

Singh, Karam Manimohon. 2006. *Nupi Lan* (Women's War of Manipur). Imphal: KPD.

8

Manipur's Economy from 1949 to 1972 : Birth and Nurturing of a Dependent Economy

Hanjabam Isworchandra Sharma

In the pre-British colonial period, Manipur had been a self-sufficient agrarian economy. Advent of colonial rule after its defeat in the Anglo-Manipuri War of 1891 brought about major structural changes in the economy subservient to the interests of the colonial power. Introduction of private property in land, monetization of economy through commercialization of agriculture and taxation strengthened the colonial grip on Manipur. Opening of the vital road links of Cachar and Dimapur served as an outlet for export of large quantities of agricultural and forest products affecting price and local consumption. It subsequently led to depletion of forest resources. Emerging economic trends witnessed the influx of cheap consumer goods manufactured in industrialized Europe and other parts of British India along with the introduction of a trading class comprising mainly of Marwaris and Bengalis. Perpetuation of these colonial economic trends played a crucial role in native deindustrialization. 'Merger' with the Indian Union in 1949 plunged Manipur's economy in the lap of a centrally planned economy. In an attempt to understand the economic undercurrents responsible for massive underdevelopment in the State, the essay critically surveys Manipur's economy from 'merger' in 1949 to statehood in 1972. Reduced to a Union Territory within the newly formed Indian Union, critical analysis in terms of economic planning that was absolutely in the hands of the bureaucrats in New Delhi is crucial. Administered and governed by agents of the Central Government, the period is crucial as dramatic transformations engineered through policies and plans crafted in New Delhi brought about an upheaval in the self-reliant economy of Manipur. Critical importance of the period is paramount as Manipur's economic destiny was shaped during these crucial years. Relegated to the status of a Union Territory after abrupt dissolution of the democratic government—duly formed after elections based on Manipur Constitution Act, 1948—the New Delhi based Union Government was in absolute charge of Manipur. The Central Government enjoyed sole and exclusive authority on the First Five Year Plan up to Fourth Five Year Plan as Manipur's budget was prepared, presented and passed only in the Parliament. It also marks the period in which restructuring of Indian economy was carried out on war footing through economic planning with maximum thrust given to the core sectors of agriculture and industry.

Manipur's Economy on the Eve of 'Merger'

On the eve of 'merger', Manipur's economy was on the verge of recovery from the shock of the Second World War. Its agrarian economy was severely affected by the War. Literally turned into a battlefield between Allied and Axis forces in the Asia Pacific theatre of the War, Manipur witnessed large scale destruction. Heavy bombing of the Manipur valley wrecked havoc on its agriculture. Most of agricultural lands were reduced to war fields and the population of livestock depleted substantially. Momentum of economic activities declined considerably during 1944 to 1948. Agriculture suffered immensely as a result of the continuing instability of the general masses, acute shortage of plough cattle, seeds and agricultural implements due to large-scale destruction during the War (Naorem, 1998, 184). Realizing the great importance of agricultural development and increasing food production for better administration, the State Government initiated a number of steps (Loitongbam, 1989, 80) as given below :

1. A separate agricultural office was established in March 1946.
2. An experimental-cum-seed multiplication farm was established at Mongjam village in 1946.
3. An experimental farm was also established at Kanglatong.

Non-Manipuri traders who fled Manipur started returning. The native traders strongly opposed return of those traders though they could not replace the efficacy and expertise in external trade. They were actually ineffective in external trade on account of lack of capital and outside contact. The wholesale dealers in Calcutta and Gauhati supplied only to the 'established businessman' who ironically turned out to be outsiders. A fraternal alliance existed among non-Manipuri traders in Imphal and wholesale traders in Calcutta and other commercial centres of India (Constantine, 1981). Instability among the indigenous trading class was largely responsible for disoriented trading pattern and creeping in of organized blackmarketing in Manipur. Shortage of essential commodities and rising prices severely affected the economy resulting in the 1948 famine. With slow down in trade and commerce and household manufacturing sector, which had already been a victim of British colonial exploitation, the economy was marred by the acceleration of land alienation. Number of sales cases instituted increased from 27,562 in 1945–46 to 38,543 cases in 1946–47. The local elite started investing their capital in land (Naorem, 1998, 185).

At the time of Manipur's 'merger' with the Indian Union, the economy had returned back to normalcy. According to 1951 Census, 83.4 percent of the population was engaged in agriculture, 5 percent in manufacturing and 4.2 percent in trade and commerce. Prior to the 'merger', Manipur was a subsistence agricultural economy with minimal industrial base in the form of handloom and handicrafts. Trading provided a substantial part of secondary employment (Chongtham, 1985, 101). In spite of its nature of subsistence production, Manipur had been a surplus exporter of rice during the colonial period. The British colonial policy of free trade made rice play a major role in the export sector leading to commercialization of agriculture. Introduction of private ownership in land traditionally inherited irrigation system and reclamation of more land for cultivation due to increased demographic pressure coupled with the trend of deindustrialization had strong impetus on the production of rice.

Land under cultivation in the Manipur valley increased from 26,500 hectares in 1891 to 75,370 hectares in 1941. Rice production in Manipur valley increased from 38,160 tonnes in 1891 to 1,08,533 tonnes (see Table 8.1). Rice surplus extracted also increased from 26 percent in 1891 to 34 percent in 1941. Rice and rice product exports increased from 2,655 tonnes in 1900–1901 is 21,694 tonnes in 1938–39 and eightfold increase during four decades (Chongtham, 2005, 152).

Table 8.1: Area Production and Rice Surplus in Manipur Valley (1891–1941)

Year	*Valley Population*	*Area (Hectare)*	*Demand (Tonnes)*	*Rice Production (Tonnes)*	*Surplus (Tonnes)*	*Percent Surplus*
1891	1,35,782	26,500	28,107	38,160	10,053	26
1901	1,91,207	42,283	39,580	60,755	21,175	35
1931	2,86,843	71,316	56,376	1,02,695	43,319	42
1941	3,43,694	75,370	71,145	1,08,533	37,388	34

Note: (i) Demand per person per annum at 207 kg. as in NSS.
(ii) Production at the rare of 40 pots per paree. Least of 75 pots (R. Brown), 43 pots (Administrative Report, 1901–02).
(iii) One pot of paddy is 54 kg *vide* R. Brown (1873: 86).
(iv) Paddy gives 2/3 rice when husked.

Source : Government of Manipur, Administrative Reports for Population and area under paddy.

The colonial period was also marked by sharp increase in valley population—from 1,35,782 in 1891 to 3,43,694 in 1941. Despite the increase in population, production of rice sufficiently met local demand well and exports requirements. However, in 1939 excessive rice export seriously affected price and local consumption leading to the Second *Nupilal* (Woman's war) against rice export. Though rice was the staple food as well as commercial crop during the British colonial rule, significant steps for improving cultivation could not be seen. For the first time, the Manipur State Durbar invited the Office of the Political Agent to take up necessary steps in bringing out an overall improvement upon the old and primitive method of cultivation in the state. Consequently, J. Shakespeare requested the help of the Director of Agriculture of Eastern Bengal and Assam in giving necessary help in this direction. Subsequently, two Manipuris were sent to Gauhati (Assam) and Dacca (Bengal) to learn improvised methods of cultivation. During First World War (1914–1919) a small group comprising mainly individuals from the hill district were selected and sent to European countries to learn different methods of cultivation under the Labour Corps Programme. Besides, in 1919 two individuals from Mao village visited France during the war to examine the possibilities of introducing terraced cultivation in the slopes of the Manipur hills (Ali, 1988, 87– 88).

These steps were largely ineffective, as they were not backed by meaningful expenditure for agricultural development. With land revenue constituting nearly 50 percent of the total revenue of Manipur (Loitongbam, 1989, 53), the colonial administration extracted huge amount of revenues from the commercialization of agriculture. However, its importance was yet to receive due attention as reflected by the colonial pattern of expenditure.

It is interesting to note that despite marginal attention, production of rice was phenomenal. It was the driving economic force behind colonial rule in Manipur. Not only was the increase in production of rice important but productivity of rice also played a crucial role. Productivity of rice in Manipur was high during the pre-colonial period. Figures indicate paddy yield of 2,700 kg per hectare in the worst land and 4,320 to 5,400 Kg of paddy per hectare on the best land (Brown, 1873). Official estimate revealed a yield of 4,959 kg per hectare on the average, which ironically is not much below the yield in post-Green Revolution of Manipur (Chongtham, 2005, 150). Even on the eve of centralized economic planning (1951), rice yield in Manipur Valley was 1,422 kg per hectare compared to all India's 714 kg per hectare (see Table 8.2).

Table 8.2: Estimates of Area, Production and Average Yield per Hectare of Rice in Manipur during 1951–1972 along with corresponding All India Average Yield per Hectare

Year	Area, '000 Hectares	Production '000 Tonnes	Yield (Manipur) Kg. per Hectare	Yield (All India) Kg. per Hectare
1951–52	81.5	115.9	1422	714
1952–53	72.0	96.4	1338	764
1955–56	90.0	131.3	1459	874
1960–61	89.0	112.2	1261	1013
1963–64	91.0	95.0	1044	1033
1970–71	140.2	159.8	1140	1125
1971–72	135.8	158.6	1168	1141

Note : The estimates for the year 1951–52 to 1963–64 are for central valley only.

Source : Government of India, Directorate of Economics and Statistics, Government of Manipur, Directorate of Economics and Statistics and Government of Manipur, Revenue Department.

In the manufacturing sector import of chief consumer goods from mainland India had a scathing effect on traditional handloom and handicraft industries. British colonial policy of free trade crippled handloom and handicraft industries, an occupation next to agriculture. Hence, major source of consumer goods used in day-to-day life came from imports that were under the exclusive monopoly of the non-indigenous Bengali and Marwari trading class. British colonial rule left and economy that was distorted and tertiarised to a certain extent; but it still possessed a strong agricultural base with sizeable surplus upon which an economy with a stronger production base could presumably be built (Chongtham, 2005, 157). Subsequent to Manipur's 'merger' with India in the post-British period, structural transformations in the economy addressing the people's aspirations were expected. Agriculture as the backbone of the economy expected to be accorded top priority along with revival of traditional handloom and handicraft industries hibernating during the colonial period. It is logical that agriculture should be given priority in line with the history of economic development of various advanced nations where agriculture was the springboard for development.

Planning for Development or Dependency

India's long journey for planned economic development began in 1951 with the inception of First Five Year Plan following the efforts for a planned economy initiated during the fag end of the colonial period. The Bombay Plan in 1944 by eight industrialists, Gandhian Plan in 1944 by Shri Sriman Narayan, People's Plan in April 1945 by M.N Roy and Sarvodaya Plan in January 1950 by Shri Jaiprakash Narayan can be regarded as the embryo of India's economic planning. Consequently, the Planning Commission came into existence on March 15, 1950. On the eve of the commencement of the First Five Year Plan the country's economy was gripped by the economic recession of the Second World War and the 1947 partition. Naturally, the First Five Year Plan accorded highest priority to agriculture as the country faced acute shortage of food and industrial raw materials. In this connection, the Planning Commission stated that agriculture including irrigation and power must have top most priority for the immediate five years period. The members were convinced that without a substantial increase in the production of food and raw materials needed for industry, it would be impossible to sustain a higher tempo of industrial development. So in the First Five Year Plan (1951–56) the Commission allocated 31 percent of public sector outlay for agricultural purposes including 15 percent for agriculture and community development, and 16 percent for irrigation and flood control. Power accounts for 13 percent, industry 6 percent, transport and communication 27 percent and social services and other 23 percent.

In marked contrast Manipur presents a different picture compared to the country's plan allocation. Agriculture in the State never received as much attention as it should. Agriculture and allied services, which include horticulture, soil conservation, forest, animal husbandry, fisheries, community development, Panchayat, and minor irrigation, received a meagre 4.06 percent of the total outlay. Industry accounts for 0.39 percent. Bulk of the total outlay (87.54 percent) was allotted to transport and communication and social and community service.

India's Second Five Year Plan (1956-61) commonly known as the Mahalanobis Plan focused on the expansion of public sector so that 'commanding heights' of the economy was under government control. It was a grand vision for propelling India on the path of industrialization. Major thrust was given to rapid industrialization. The Planning Commission stated that rapid industrialization and diversification of the economy constituted the core of development. The country, therefore, needed to develop basic industries, particularly those industries that manufacture machines needed for further development. Subsequently, industries were allotted 24 percent of total outlay. Even though priority was given to industry, agriculture was not neglected receiving 30 percent of the outlay.

The plan outlay for Manipur was in complete disjunction with the overall trends in the country as only 2.08 percent of the total outlay was allotted for industries. Agriculture and allied activities received 17.52 percent. Like the previous plan, bulk of the outlay went to transport and communication and social community services with 71.16 percent.

Commencement of the Third Plan (1961-67) witnessed the beginning of long term perspective planning to achieve the objective of an integrated and balanced growth of industry and agriculture. Agriculture including community development and irrigation was allotted 23 percent of actual plan outlay and industry 24 percent. Power was allotted 14.6 percent and transport and communication received 24.6 percent. Manipur continued its trend of disjunction and disorientation with agriculture and allied services getting 14.83 percent of total outlay, industry 3.87 percent and power 9.19 percent. Illogical and irrational emphasis on transport and communication and social services still continued with 70.59 percent.

Delay in the Fourth Plan due to hostilities with China and Pakistan resulted in three Annual Plans from 1966 to 1969. Even during this period the all India trend underlined agriculture as high priority with 24 percent of the outlay. Considered the beginning of Green Revolution in India, the period is marked by large-scale introduction of high yielding (HYV) crops along with proportionate application of fertilizers and irrigation. However, the plan outlay of Manipur in this period was not significantly different from the previous plan. Agriculture and allied services received 12.43 percent and industry 4.34 percent.

In the Fourth Plan (1969-74) agriculture, power and industry were given due attention, getting 24 percent, 15.4 percent and 22.8 percent of plan outlay, respectively. It was the last plan in the pre-statehood period of Manipur. Agriculture and allied activities got 11.78 percent, industries 3.88 and power 14.05. Like earlier plans, maximum outlay went to transport and communication and social services with a staggering 67.35 percent. The conclusion is obvious—the country's plan allocation was carried out in such a manner that special emphasis was paid to the more important sector of agriculture and industries without compromising other sectors. Irrespective of plan achievement different sectors were paid due and proportionate attention. However, Manipur presented an immensely different picture. Disproportionately high priority was given to transport and communication and social and community services during the first four Five Year Plan periods including *ad hoc* plan periods with the State plan outlays greatly oriented towards these two sectors.

Vital sectors such as agriculture, irrigation, power and industries were conspicuously neglected up to the end of the second decade of planning. Thus the two sectors accounted for 91 percent, 73 percent, 77 percent, and 66 percent of the total outlay in the First, Second, Third, and Fourth Plan of Manipur, respectively. Juxtaposition with the all India pattern of 50 percent, 46 percent, 20 percent and 38 percent in the same sectors reveals the lopsided and disproportionate nature of Manipur's plan outlays. In a State where 90 percent of the area is hills, concern for transport and communication is understandable. However, no stretch of imagination can reconcile the grossly disproportionate investment on the transport and communication sectors to the ruinous neglect of agriculture (Chongtham, 1985, 100).

Perpetuation of Manipur's status in the post British period as a peripheral economy is quite evident from the pattern of plan outlays of the State. Marginalization of agriculture and industry compelled the State to depend heavily on imports from the core economy of mainland India right from the initial stage of planning. Manipur was steadily transformed into market. Agriculture, the mainstay of economy with 80 percent of the population engaged in it suffered immensely as a result of sheer neglect. The Five Year Plans had a sabotaging effect on the agrarian economy of the State as it increasingly became dependent. Agricultural production became stagnant. Annual production of rice during 1951-1971 was an insignificant 0.2 percent as against the all India average of 1.2 percent (Chongtham, 1985, 100). The pre-Statehood period also witnessed a surging tide of influx of immigrants creating intense demographic pressure. Decennial growth rate of population was 35.04 percent in 1951–61 and 37.53 percent in 1961–71 while it was only 12.80 percent in 1941–51 (see Table 8.3). Local production of foodgrains particularly rice was not able to meet the demand for local consumption of the rapidly increasing population. Manipur was compelled to rely heavily on the import of edible items such as rice, wheat, pulses, oil, etc. from mainland India.

Table 8.3 : Decennial Growth Rate of Population (1941–71)

Decade	*Decennial Growth Rate (%)*	
	Manipur	*India*
1941–51	12.8	13.31
1951–61	35.04	21.51
1961–71	37.53	24.8

Source : 1. Government of India, 1961 Census of India: Manipur. New Delhi: Office of the Registrar General, India.

2. Government of Manipur, various years. Imphal: Directorate of Economics & Statistics.

The period under study especially the late 1960s was marked by great transition in Indian Agriculture. Green Revolution through the introduction of HYV seeds, chemical fertilizers and proper irrigation in agriculture was initiated in 1968–69. Its impact was so momentous that the Green Revolution belt viz. Punjab, Haryana and western Uttar Pradesh started producing greater surplus output. Manipur remained aloof despite introduction of HYV seeds much before the onset of Green Revolution in India. Importance of irrigation was conspicuously felt only in 1972 when the State was hit by a severe drought. The government had never taken irrigation seriously (Thingujam, 1998, 29). There was substantial increase in the Second and Third Plan outlay for agriculture and allied activities, but irrigation never received due attention. It was only in the Fourth Plan during 1973–74 that the Loktak Lift Irrigation Project, the first major irrigation project in the State, was taken up meagre plan investment of Rs. 1.41 crore (Thingujam, 1998, 35).

Lack of proper irrigation was a major hurdle for crop diversification in Manipur. Rice was the only important *kharif* crop and covered 80 percent of total cultivable land. Scope for diversification of important *rabi* crops such as wheat was extremely limited in the State in the absence of irrigation. Area under wheat remained confined to certain marginal pockets only (Loitongbam, 1988, 138). With increasing demographic pressure and neglect of agriculture, self-sufficiency in the agriculture especially foodgrains declined rapidly. Agriculture during the first four Five Year Plans was severely affected mainly because of the failure to take up development schemes in this sector (Loitongbam, 1988, 67).

Right from the colonial period the State increasingly became dependent on mainland Indian economy for consumer goods. The trend continues even after 'merger' with India. Industry was kept in the back in the Five Year Plan prior to statehood. Centralised planning never really initiated any move towards development of medium and large-scale industries (Singh, 2005). Plan outlays for the first Three Plans were exclusively meant for village and small-scale industries. It was not satisfactory as handloom and handicraft industries constituted an important category next to agriculture in terms of generating employment and income. Meagre allotment in the plan outlay resulted in thwarting the potentials of this sector. Attention was given as late as 1976 with establishment of Manipur Handloom and Handicraft Development Cooperation. Large and medium scale industries were paid attention only during the Fourth Plan with the establishment of Khandsari Sugar Factory in 1973.

When Manipur was struggling to lay down an industrial base even for village and small-scale industries, the country had already consolidated its strong industrial base for heavy and basic industries as well as village and small-scale industries. During the Second Plan a major task in industry was the building up of three steel plants in the public sector: Rourkela Steel Plant in Orrisa, Bhillai Steel Plant in Madhya Pradesh and Durgapur Steel Plant in West Bengal. Other programmes of industrial development included manufacture of electrical equipments, expansion of Hindustan Machine Tools, expansion of Sindri Pesticides Factory, establishment of a fertilizer plant at Nangal and expansion of Hindustan Shipyard and Chittaranjan Locomotive Factory. There was also rapid expansion of machine-building industries for use in agriculture, transport and industries like chemicals, textiles, jute, cement, tea, sugar, flour and oil mills, paper and mining, etc. Good progress was also recorded in modernization and re-equipment of important industries such as jute, cotton, textiles and sugar. Quite a number of new industrial items like industrial boilers, mill machines, tractors, motorcycles, and scooters were also produced in large quantities. In the sphere of village and small industries, substantial progress was recorded. About 60 industrial estates comprising 1,000 small factories were set up and production of number of items, such as machine tools, sewing machines, electric motors, fans, bicycles, hand tools, etc. increased from 25 to 50 percent during the five year period. Khadi, handloom and powerloom cloth production increased from 1,610 million metres to 2,150 million metres (Datt and Sundaram, 1999). The period was also characterized by a policy of import substitution; the share of consumer goods, which were imported, was already low at the beginning of 1950s. By the end of 1960, India imported practically no consumer goods (Sodersten, 1984). On the other hand, Manipur became increasingly dependent on consumer goods from mainland India.

The pattern of Plan investment in the State carried out during the pre-Statehood clearly depicts a colonial hangover. Overemphasis on transport, communication and social services in the initial stage of the planning was a legacy perpetuated since British colonial period.

A cursory glance at Manipur's actual expenditure in 1936–37 reflects similar investment patterns. Public works, education and medical services accounted 35.3 percent of the total expenditure.

During British rule some inter-village and inter-State roads came up to subserve colonial interest which helped little in development of the native economy since important sectors like agriculture and industry never received due attention. Few steps were taken up to improve household and cottage industries by instituting the Department of Arts and Crafts. Expenditure under this head was about 1.2 percent of the total expenditure. Pattern of expenditure during British rule never intended any meaningful effort toward development (Ali, 1988, 229). It continued even after 'merger' with India right upto the attainment of statehood. Transport, communication and social and community services were allotted 69.5 percent of the overall plan outlay amounting to Rs. 6,105.38 lakh during the first four Five Year Plans. Even though transport and communication amounted to plan expenditure of 72 percent, 35 percent, 40 percent and 40 percent for the first four-Five Year Plans, respectively, there was no remarkable development in this area. Road density stood at a meagre 12 km per 100 sq. km as against 29 km at the all India average in the year 1971. Length of surfaced road was just 17,652 km accounting 62 percent of the road length, ironically inclusive of a road length of 208.8 km National Highways. Village road constituted 40 percent (1,132.2 km) of the road length. Even in 1971, 90 percent area of the State remained without roads (Keisham, 1994).

The policy of neglecting the productive base of economy, inherited from the British policy of treatment of the State as part of a 'frontier', still continues. It connotes only strategic importance from the defence matrix of India (Konthoujam, 2006). Economic dependency is seen as an instrument for strengthening the integration of the State to the Indian Union. It is further accentuated by huge influx of non-Manipuri population from mainland India, Bangladesh and Nepal. In August 1946 and July 1947, the Governor of Assam instructed the Political Agent to allow re-entrance of 18 and 19 important *Marwaris* traders respectively (Naorem, 1998, 83).

By the end of the year 1947, the *Marwaris* in Manipur numbered roughly 2,000 (Salam, 1994) and ever since, the number of outsiders has been increasing day by day in the post-British period.

Heavy dependence for agricultural products and consumer goods met through external trade under the monopoly of non-Manipur traders virtually robs the State of any meaningful reinvestment as huge profits are siphoned outside. Overemphasis on transport and communication in centralised planning only subserves the interest of external trade entirely dependent on the non-Manipur traders. They have emerged as an important economic catalyst between mainland India and Manipur.

The notion of economic backwardness of the State due to geographical isolation and landlockedness does not hold ground, as the productive base of the economy was never given a chance to flourish. The seed for a dependent economy was already sown in the pre-Statehood period. Manipur was undeveloped on the eve of 'merger' but its underdevelopment started with the 'merger' in the Indian Union. The Union Government's claim that Manipur was suffering from underdevelopment before 'merger' does not contain the whole truth. Its economy was in a satisfactory state to 1949. Majority of the Manipuris unanimously agree that it has been facing steep decline since the 'merger' (Paratt, 2005, 119).

REFERENCES

Ali, Md. Ashraf. 1988. *Economic history of Manipur, 1801-1951*. Ph.D. Thesis, Manipur University.

Annual Administrative Report (Manipur), 1991–20 and 1936–37.

Brown, R. 1873. *Statistical Account of the Native State of Manipur and the Hill Territory under its Rule*. Calcutta: Office of the Superintendent of Government Printing.

Chongtham, Priyoranjan. 1985. Aspects of the State Economy and Planning for 2000 AD. Paper presented at Manipur 2000 AD, October 11–13, in Manipur University, Manipur.

——. 2005. Manipur's economy: historical roots and structural evolution. *Eastern Quarterly* 3, III. October–December: 149–59.

Constantine, R. 1981. *Manipur: Maid of the Mountains*. New Delhi: Lancer Publishers.

Datt, Rudder and K.P.M Sundaram. 1999. *Indian Economy*. New Delhi: S. Chand & Company.

Frank, Andre Gunder. 2001. The development of underdevelopment. In *Imperialisms: Critical Concepts in Historical Studies*. Vol. II, ed. Peter J. Cain and Mark Harrison. London and New York: Routledge.

Government of India. Various years. *First five year plan, second five year plan, third five year plan, annual plan and fourth five year plan*. New Delhi: Planning Commission.

Keisham Manibabu. 1994. Financing of rural development in Manipur 1988–90. Ph.D. Thesis, Manipur University, Manipur.

Konthoujam, Indrakumar. 2006. Imagining the North East through India's Look East Policy: Towards a contextual understanding. *Alternative Frames* I, II. January–March: 45–60.

Loitongbam, Udhob. 1989. *A Brief History of Agricultural Development in Manipur*. Imphal: OK Book Store.

Meitei, K. Ibomcha. 1991. A study on the Khadi and village industries in Manipur. Ph.D. Thesis. Manipur University. Manipur.

Naorem, Lokendro. 1998. *Unquiet Valley*. New Delhi: Mittal Publication.

Paratt, John. 2005. *Wounded Land: Politics and Identity in Modern Manipur* New Delhi: Mittal Publication.

Salam, Irene. 1994. The Marwari community and the economy of Manipur during the period of British paramountcy 1891–1947. Paper presented, Our Freedom, December 12, in Imphal, Manipur.

Singh, M. Iboton. 2005. Political economy of Manipur: Transformation within the mixed economy paradigm. *Eastern Quarterly* 3, III. October–December : 170–82.

Soderstern, Bo. 1984. *International Economics*. London: McMillan Publishers.

Thingujam, Roben. 1998. Economics of minor irrigation in the three valley districts of Manipur. Ph.D. Thesis, Manipur University, Manipur.

9

Surface Transport and Economic Development in the Manipur State

Yambem Imo Singh

Introduction

With the integration of the erstwhile kingdom of Manipur with the newly independent Indian Union in 1949 in the wake of British departure from the subcontinent, it came under great influenced by the socio-economic and political developments crafted in the power corridors of distant New Delhi. Manipur has been connected with the national road networks through the national highways and with the advent of modern modes of transport such as lorries and trucks it became a part of the national market system. Thus linked with the rest of the country mainly through NH-39 and NH-53, it receives a variety of commodities from the outside markets and in return exports her local produce which are of a primary nature at an insignificant level. However, the condition of roads and surface topography do not allow speedy intra- and inter-State movements of people and commodities. Rather they increase the friction of distance. Acute lack of modernization and improvement in the narrow and constricted transport routes in the hills has put the State in adverse positions. Primitive conditions of the highways and roadways make them vulnerable to seasonal influences. Heavy rains during the summer season also pose problems in the inter-regional transport system.

Over the years the State has visualised marginal growth in the transport network, particularly roads. When the First Five Year Plan was launched the length of roads increased to 570 km including 216 km of NH-39. During 1951–1961, the New Cachar Road was improved and upgraded to NH-53. The public works department has been in charge of improvement and maintenance of NH-39 for a length of 188 km from Maram to Moreh since statehood in 1972. The portion from Maram to Pallel has been widened to double lane. The length of National Highways in the State, between 1986 and 1996 grew from 434 km to 438 km The figures indicate a dismal picture where the growth of national highways in the North-Eastern region is far below the all-India average. All the States of the region share less than 2 percent of the all-India total.

Evolution of Transport

Evolution of transport in the State of Manipur can be studied under three main periods, viz. (1) Ancient Period; (2) Medieval Period; and (3) Modern Period. Although it is difficult to ascertain the exact age of settlement and advent of Manipuri civilization, there is evidence of human settlement in Manipur since the sixth or even eighth century B.C. (Dun, 1886). Around second century B.C. there was regular trade through a route which connected China, Afghanistan and Europe *via* Manipur, Assam and Upper India. In his history of Burma, Harvey (1929, 8) maintains that upper Burma lay inaccessible though it was nearer to China. It changed during the second century B.C. with the emergence of trade routes through Burma—two were along the Irrawaddy and Saleveen River and the third down the Chindwin River through Manipur, which took caravans about three months to reach Afghanistan where the silk of China was exchanged for the gold of Europe. In his *History of South East Asia*, Hall (1955) also mentioned about a road connecting lower Burma with India *via* Irrawaddy, Chindwin and Manipur. There were certain valley routes connecting the Brahmaputra valley with the Kabow valley. Table 9.1 below shows some of the important hill routes connecting various countries with Manipur.

The hill routes converge at Imphal and they again diverge in different directions so as to link with different places in Burma in the east and south, and Cachar and Assam in the west. The people throughout the ancient and middle ages used these hill routes.

Table 9.1: Important hill routes connecting various countries with Manipur

	Name of the routes	*Countries and places connected*
A.	1. Heirok Route *via* Machee 2. Aimol Route 3. Ngara Chingjin Route 4. Maring Route	Kabow, Ava (Burma), China and Sio South-Eastern Countries of Asia
B.	1. Tongjei Maril Routes (Old Cachar Road) 2. Ngaprum Chingjin Route 3. Khongjai Route 4. Acqui Route	Cachar, Tripura, Sylhet and other western countries
C.	1. North Eastern Route (Tekhao Route)	Nagaland, Assam and other north-western area

Source: Sharma, Nandalal. 1960. *Meitrabak*. Imphal: O.K. Book Store.

Much later, in addition to the existing routes, succeeding kings constructed few more roads. King Ayangba (821–910 A.D.) constructed a road, which served the people for centuries. It still exists as the Ayangpalli Road, which is the Eastern boundary of Imphal Municipality.

Land transport was the only major means of transport for movement of people and goods. Most of the hill routes and roads in the valley were impassable. Besides, the hill routes were too narrow and steep at a number of places. The main traffic in these routes consisted of silk textiles and silk goods, bamboo and cane products, gold and animal horns. Their commodities were carried to the western countries as well as to Burma and China. Inside the kingdom, there was little trade and commerce among the principalities (Sharma, 1989). During the medieval period (1074–1830 A.D.) successive kings constructed roads. In the early sixteenth century, King Kabomba opened a new route to Assam. In about 1630, King Khagemba improved roads for better transportation of goods and people. King Paikhomba improved road from Chinga to Mongsangei in 1073 (Singh, 1965). He maintained trade relations with Burma, Cachar and Tripura (Sharma, 1989). The important roads, which connected Manipur with Sylhet and Cachar district of Assam, were the Aree Route, the Kala Naga Route and the Khongjai Route (Pemberton, 1966). These routes were the major routes for trade and commerce and movement of people. Other important routes, which connected Manipur with the Kabow valley in Burma, were the Machi Route and the Aimol Route (Pemberton, 1966). There was another line of communication that could lead to the Kabow valley. A number of routes also spread from Tamu to reach Ningthi (Chindwin). Besides passenger traffic, there were movements of military troops in these hill routes.

River transport was also common. River transport using small country boats also helped in moving goods and people. Boats were used for movements of passengers and goods in the Imphal and Iril rivers. Bigger rafts and boats were used in the Barak and Jiri rivers. Merchants traded in silk and various forest products through these rivers.

Transport development in the modern period began around 1830s with the advent of the British domination over the kingdom of Manipur. Construction of roads in the hill areas became an important feature. Construction of Cachar road was taken up in 1837 and completed in 1844 (Singh, 1965). In 1881, the Imphal Dimapur Road was completed and became much more important than the Cachar Road. During the nineteenth century four important roads in the west (Dun, 1886):

1. The Kohima Road (Imphal-Dimapur Road *via* Mao and Kohima),
2. The Rozepemah and Kenom Road,
3. The Cachar Road *via* Khoupum valley and Bishnupur, and
4. The Acqui Road as the Upper Cachar Road came up.

Important hill routes besides innumerable routes like the Aimol Route, the road leading to Kabow valley *via* Haitup Pokpi and the road from Manipur to Moreh *via* Pallel leading to Kabow valley also emerged. About ten main roads were recognized to have radiated from Imphal. These roads were not metalled roads but they were wide and levelled roads.

National Highways 39 and 53

As in other States of the North East, roadways are important means of communication because they connect even isolated villages over the hills. Road transport in the State continues to play an important role in carrying commodities and people. This has acquired great significance for short distance movement. The only important link between Manipur and the rest of the country is the Imphal-Dimapur Road that connects Imphal with the railhead at Dimapur. This highway (NH 39) passes through the Naga Hills in order to connect Manipur with the Brahmaputra valley. It passes through centres like Mao, Maram, Senapati, Kangpokpi, Sekmai, Khonghampat and finally reaches Imphal city. From Imphal this highway proceeds towards Moreh, a border town located on the Indo-Myanmar border, passing through centres such as Thoubal, Khongjom, Pallel, Tengnoupal and Khudengtabi. It traverses five districts—Senapati, Imphal East, Imphal West, Thoubal and Chandel. Similarly National Highway 53 joins Imphal City passing through centres like Jiribam, located on the extreme west, Nungba and Noney. The road is also called the New Cachar Road having 225 km between Imphal city and Jiribam and constitutes a second line of communication between Manipur and the rest of India. This National Highway runs from Badarpur *via* Silchar and Jirighat to Imphal.

With the exception of a small portion of the roads in the plain areas, most of the roads in Manipur run winding and twisting with very steep elevations. Existing landforms play an important role in the construction of the roads The National Highway 39 enters Manipur from the north through Nagaland at Mao at an elevation at about 2000 m and after bending and twisting it descends near Kangpokpi at about 1500 m. It then comes down with gentle gradient upto Sekmai with an elevation of about 790 m. From there it straightens its way to Imphal and with few curves its straightness continues upto Pallel in the south. From there this road ascends steeply with many curves in the mountainous area upto Tengnoupal where the elevation attains roughly 2000 m. From there it climbs down to Moreh at about 200 m. Similarly, the National Highway 53 enters Manipur at Jiribam at about 60 m. After running few kilometres on the Jiri plain, it ascends gradually on the steep mountains. It also crosses many rivers such as Makru, Barak, Irang etc. It is about 224 km from Imphal to Jiribam. This road, in spite of its being a National Highway, is too narrow for vehicles to cross. Besides this, it is constructed on the high and steep slopes of hill and mountains. The bridges over the rivers such as Makru, Irang and Barak are also too narrow and weak for heavy vehicles. Similar features exist in other State highways such as the Tiddim Road, Imphal-Ukhrul Road, Imphal-Sagolmang Road, Imphal-Kangchup Road, Imphal-Yairipok Road, Imphal-Mayang Imphal-Sugnu Road etc. (Sharma, 1989). The two highways serve the State as two lifelines through which Manipur is connected with the neighbouring States of Assam, Nagaland and Mizoram. They are supposed to be all-weather motorable routes. However, they are frequently blocked and become impassable during the rainy reason due to landslides. They are also prone to economic blockades, which have serious economic setbacks for the State. The total length of National Highways in Manipur is 434 km.

Accessibility

Accessibility still remains an important issue in Manipur even in the twenty-first century. There are areas that are not easily accessible by roads. Similarly there are places that are least accessible. About 90 percent of the total geographical area of the State is hilly terrain where accessibility is difficult even during dry seasons due to lack of good roads. Moreover, acute lack of road connectivity still makes many parts of the State highly inaccessible.

Accessibility by road varies from one region of the State to the other. For example, in the eastern region, which is delimited by the Imphal-Ukhrul Road on the west and a part of Imphal-Moreh Road on the south, areas beyond 16 km are inaccessible all the year round and comparatively have the highest inaccessibility in the entire State (Sharma, 1989). Similarly, about 30 percent of the area in the south, which comprises the southernmost tracts of Churachandpur and Chandel districts, remains inaccessible all the year round. Similarly, the south-western region lying between Imphal-Jiribam Road (NH 53) on the north and Imphal-Tipaimukh Road on the south and the north-western region comprising the districts of the Tamenglong and Senapati have lots of inaccessible areas. For the northern region, which comprises the area delimited by the Imphal-Dimapur Road on the west and Ukhrul-Kharsom-Jesami Road on the east, most of the areas are accessible. Accessibility is highest in the central region, which comprises the plain districts of Imphal, Thoubal and Bishnupur. Inaccessibility is highest in the eastern region of the State (Table 9.2). Inaccessible areas are characterized by precipitous hilly terrain with thick forest cover. These areas receive heavy rains. This feature is absent in the central region where most of the nodal points are well connected by one or more surfaced roads.

Table 9.2 : Road Accessibility in Manipur

Region/District	*Approximate areas beyond 16 km are inaccessible*
Eastern	1,984 (41.3%)
Southern	1,472 (30.3%)
South-Western	768 (20.3%)
North-Western	448 (10.1%)
Northern	–
Central	–
Manipur	4672 (20.9%)

Source: Sharma, B. Lalhari. 1989. *Transport Network in Manipur: A Geographical Analysis.* Ph.D. thesis, Manipur University, Manipur.

Accessibility in the hill areas seems to be more variable because of the fact that the terrain conditions do not provide unevenly spaced settlement owing to the undulating topography. Both primitive and modern types of transportation interconnect these settlements. Roads provide better links to these settlements. Some of the hill settlements are not reached even in two days walk from the nearest surfaced road. In general, hill economy remains backward due to these factors. Accessibility in the State is low and desperately needs to be improved by rapid construction of new roads especially in the hills so as to make them accessible. Balanced economic growth in various regions will emerge only when there is greater nodal connectivity and improved transportation network.

Road Traffic

Road transport in the State is used for both passengers and goods traffic in the State. This is because of the fact that the physiography allows only road transport in Manipur among various modes of surface transport. Road transport solely carries intra- and inter-State transportation of both passenger and goods. From this point of view the two main National Highways 39 and 53 are the two most important links of the State. There are a number of transport agencies that carry both passengers and freight traffic internally as well as externally along these highways and other State highways. Before the advent of the government transport agency, the Manipur State Transport (MST) later converted into the Manipur State Road Transport Corporation (MSRTC) served as an important transport operator internally as well as externally upto Dimapur during the 1930s and 1940s. After 1976, the corporation operated both passenger and freight traffic in and outside Manipur. It later took the responsibility of inter State passenger traffic. In the 1980s private agencies emerged to share transport responsibility. In case of goods, agencies like M.D.U., Manipur Apex Co-operative Marketing Society, Food and Civil Supply Department, Public Works Department and various private operators like ABC India Limited, Om Carriers, Speedy Carriers, Zet Travels and North East Carriers are operating their fleet of buses and trucks in the movement of intra- and inter-State goods traffic. Imphal being the central route serves as both terminating and originating place for all traffic-goods and passengers while other small town district headquarters and administrative centres serve as originating and terminating places for different routes.

Local Markets and Variation in the Density of Traffic Flow

Existence of local markets influences the variation in the density of passenger and goods traffic flows among other factors. In the case of Indo-Burma Road, important local daily markets are Lilong, Thoubal, Yairipok, Wangjing, Pallel, Chandel, Kakching, Sugnu and some other smaller daily markets. On Tiddim Road important markets are Nambol, Oinam, Bishnupur, Ningthoukhong, Moirang, Kumbi, Wangoo and Churachandpur. On the Imphal-Dimapur Road Sekmai, Motbung, Saparmeina, Kangpokpi, Senapati, Maram and Mao are important daily markets. On the other hand, low traffic density is mainly due to the absence of such markets. For example, Kangchup and Lamlong on the Imphal-Kangchup Road, Samurou, Mayang, Imphal and Wangoi on Imphal-Mayang-Imphal Road, Lamlai, Litan and Ukhrul on Imphal-Ukhrul Road play important roles in traffic intensity. It is observed that the larger number of markets along the Indo-Myanmar Road and Imphal Dimapur also indicate the economic importance of these roads. These sections of NH 39 contribute significantly in terms of the economy of the State and perhaps it is one of the two economically viable routes of the State yet to be exploited fully.

Goods Traffic

In the mid 1980s there were three types of services under traffic operation, viz. (i) Inter-State service from Imphal to Dimapur and from Imphal to Silchar *via* Jiribam, (ii) District services; and (iii) City services. During this period there were 39 routes under district services and 10 routes under city services. At present, there is no centralised agency for goods transport. It is entirely operated by private operators. There are various agencies operating in Imphal on goods traffic like ABC (Assam Bengal Carrier) which is the biggest agency with largest number of trucks on the roads. There are other agencies operating with lesser number of trucks to carry different commodities. These agencies bring goods from outside into the State. There is a heavy concentration of goods traffic on the Imphal-Dimapur Road with a flow of about 130 to 150 trucks per day carrying different commodities. Goods ranging from foodgrains to fuel and textile goods are transported from neighbouring States through NH 39 and NH 53, while the forest produces are brought from neighbouring hill districts to Imphal and other centres. Commodities, which were received through National Highway 53 as inflow in 1995, were only 3.96 percent of what was received through National Highway 39 in the same year. Outflow through this route was 2.81 percent of what Imphal centre sent out as out flow in that year through the National Highway 39.

The two National Highways traverse the State from the north to south-east (NH 39) and from west to the centre (NH 53) on which the volumes of goods traffic are higher. The two plain districts of Thoubal and Bishnupur have prosperous agricultural and piscicultural economy. Therefore, the goods traffic on the roads in these districts is mainly agricultural produces. In hill districts, forest produces comprise the majority of goods under transport. A comparatively better network of roads than the hill districts, which present a pathetic picture, serves the densely populated districts of Imphal, Thoubal and Bishnupur. Hence, the movement of goods is quite high in the densely populated plain districts. Urban centres are mostly located in these plain districts. Since urban centres are nodal points of transport, movement of goods is mostly concentrated on those roads that connect the nodal points.

As for rural transport in Manipur it is far from satisfactory. The chief goods traffic consists of agricultural commodities and its flow pattern is variable. The volume of traffic is low throughout the year, which is mostly influenced by the existing physiographical conditions as well as road conditions. Bullock carts are the only means of movement for the agricultural products to the village markets. If there are good link roads for both passenger and goods vehicles, the agricultural commodities may reach the markets more quickly. Considering these facts, there is greater need for better link roads with improved surface conditions to gear the rural economy for quick socio-economic transformation of the countryside. Efforts have to be put in to connect all the villages by good inter-village roads with the neighbouring market and service centres.

Transport Regions

Transportation has been an essential element in delineating various regions in Manipur. It can be discussed under three primary transport regions (Sharma, 1989, 19):

1. The Central Region,
2. The Eastern Region, and
3. The Western Region.

The Central Region

This region comprises the plain districts of Imphal, Thoubal and Bishnupur. Imphal is the biggest focal point for both this region and the whole of the State. There are still lots of inaccessible areas particularly in the eastern and western hills, while the plain areas are accessible all the year round despite the immensely obsolete conditions of many roads. Passenger and goods constitute the main traffic. Among the goods traffic chief items are vegetables, rice, potatoes, flour, grains and pulses, fish, fuel wood, bamboo shoots, sugarcane etc. Out of 32 towns in Manipur, 24 towns are located in this region. Many of the roads are metalled roads. Maximum traffic is concentrated towards Imphal, which is the hub of all economic activities, and diverges from it to different towns and district headquarters that are nodal points where traffic converges from different directions. Main economic activities include agriculture, industry, horticulture and pisciculture.

The Eastern Region

This region comprises the districts of Ukhrul and Chandel. This is the least accessible area of the State. In the Ukhrul region, Imphal-Ukhrul Road is the most important road. The Chandel region is served by the National Highway No. 39 upto Moreh. Villages in most parts of the region are not connected by any kind of road and thus the area remains backward economically. Half of the region is inaccessible. The main goods traffic consists of vegetables, handloom clothes, rice potatoes, timber and minerals particularly limestone. Major economic activities include mining and forestry.

The Western Region

This region comprises larger areas of Senapati, Tamenglong, Churachandpur districts and Jiribam sub-division of Imphal district. About one-third of the area lies in inaccessible areas. Because of dense forest cover, construction of new roads is immensely difficult. The northern part of the region is more accessible than other parts because of its location on the National Highway No. 39. Goods traffic include agricultural, horticultural and forest products. Rural transport away from the main roads and highways becomes difficult due to poor road conditions and steep sided hills. Accessibility is higher in Jiribam area where freight as well as passenger traffic sometimes pass through Barak and Jiri rivers. In this region the main economic activities are agriculture, horticulture and forestry.

Road Development

Under the 20 years road plan beginning with the Seventh Five Year Plan as the first phase involving construction of an additional length of 1785 km, attempts were made towards extensive widening, lengthening, strengthening and improvement of the road—State highways, major district roads and inter-village roads. There is no doubt that most part of the State especially the hill districts are highly inaccessible due to lack of metalled and all-weather roads.

Imphal as the Interacting Centre

Imphal, which acts as both the origin and destination of goods, has interacted with a number of centres located in the neighbouring States in the North East as well as other centres outside the region. It has interacted with these centres by way of surface transportation for various commodities in an increasing manner. Some of the important centres interacting with Imphal are Dimapur, Guwahati, Jorshil, Digboi, Namrup, Dhubai, Bongaigaon, Golaghat, Siliguri,

Shillong, Jagiroad, Dibrugarh, Morijan, Dhulianjan, Sibasagar, Sonari, Baorghat, Kamrup, Umrangro, Kalima, Golapur, Kolkata, Delhi, Jamshedpur, Pune, Aurangabad, Bangalore, Chennai, Rajamundri, Ludhiana, Madurai, Hyderabad, Patna, Midnapur, Ahmedabad, Panipat, Bhubaneshwar, Jaipur, Chandigarh, Muzaffarpur, Baroda, Sattur, Indore etc.

Commodity Flow and Transport Linkage

Imphal and other centres located in the State depend upon other centres for a wide range of commodities. The field of interaction has to expand well beyond regional limits. Due to lack of production base and technology, the State has to import huge quantities of agricultural products, mineral products, manufactured commodities and construction materials from other States. Similarly, domestic products, which are basically of a primary nature, are also supplied from Imphal to other centres located outside the State and the region. Imphal centre has experienced greater dependency on other centres on almost everything ranging from food items, construction material, raw materials and other commodities exhibiting an increasing trend of dependency. Commodities are brought in the State through NH 53, which is insignificant in comparison to the inflow through NH 39. Commodities entering the State can be broadly classified under the following heads (Singh, 2006) :

1. Agricultural products,
2. Mineral products,
3. Manufactured products (edible and consumable), and
4. Construction materials.

In 1995 the State registered a total commodity inflow of 1,15,940 tonnes of which 38.1 percent was agricultural products, 1.7 percent mineral products, 14.3 percent manufactured products (edible and consumable), 2.8 percent manufactured products (non-edible), 16.7 percent construction materials, 25.2 percent mixed goods and 1.2 percent miscellaneous goods. It was revealed that Imphal city was heavily dependent upon centres located in Assam,

Nagaland and Meghalaya for certain commodities. Its dependence upon the centres of Kolkata and Delhi for a variety of goods was less in comparison to centres such as Dimapur and Guwahati located closer to Imphal centre.

The distribution of the incoming goods and commodities in the State reveals two distinct patterns. First, the arrival of commodities tends to decrease with distance. Second, the arrival of commodities tends to specialise with distance. While commodities in large quantities reach here from neighbouring areas, certain items in limited quantities arrive from distant centres. Besides, seasonal trends of inflow of commodities also reveal favourable transport in winter season. Imphal centre received 60,110 tonnes of goods in winter as against 55,830 tonnes of goods in summer 1995. This was because of the summer rain that hindered transport movements in the State. The State has limited resources. Its volume of outflow was less than the volume of inflow. The State recorded about 5,160 tonnes of commodities as outflow in 1995. Commodities included agricultural products, animal products, mineral products, and forest products, manufactured products (edible and non-edible), construction materials and mixed goods. There were more than 20 centres outside the State receiving commodities from the Imphal centre. In 1995 centres like Dimapur and Guwahati received 1630 tonnes and 2800 tonnes of goods respectively. Major centres receiving commodities beyond the North East were Delhi and Kolkata. Commodities which were moved out of the State consisted of ginger, iron scrap, teak, gram, old plastic materials, empty bottles, garlic, kidney beans, clove, cement, chilies, pulses, black pepper, empty lines, old paper, wheat bran, cassia bark, chromites, handloom products, betel nut, turmeric, footwear, brooms, defective goods, groundnut, rice, hides and skin, common plywood, empty bags, black board, hard wood varieties etc. Of the commodities only a few may be selected in terms of substantial quantities sent to different centres outside the State. A crucial feature is the economic equations of these outflows. Whether they are economically viable or not needs to be studied critically. The transport linkage established along the National Highway 53 was found to be considerably weak (Singh, 2006).

The Border Town Moreh

Moreh town is situated on the NH 39 on the border of Myanmar 190 km away from Imphal. It has become an important trade centre where a number of commodities are exchanged between India and Myanmar. Major items entering India are blankets, betel nut, pulses, teak, ground nuts, iron scraps, brass scraps, gold, silver etc. Main items from India entering Myanmar through Moreh are cotton yarn, food products, cycles, textile, kerosene, diesel, pharmaceutical drugs etc. It may be observed that iron and steel scraps, electronic goods and blankets are third country products, the export and import methods such as counter trade, transit trade are important trading features in the Moreh Township.

For several decades, the two neighbouring countries had neither trade nor other kinds of formal economic relations till 1992. However the immense potential for mutually beneficial economic partnership came into focus when India and Myanmar signed a border trade agreement in 1994. It came into effect from April 12, 1995 when Indo-Myanmar border was officially declared open for trade at Moreh in Manipur. Other border trade points likely to be opened are Champai in Mizoram and Lungwa in Nagaland. The trade agreement, besides claims to improve India's market share in Myanmar, aims at strengthening the economies of North-Eastern States by establishing export oriented industrial units. Potential for either investment or exports to Myanmar have been identified for the States of the region (IIFT Report, 1995). For the States of the region items such as fruits and vegetables, tea, rubber, medicinal plants and herbs, extracts of resin from pine trees, handloom products and silk are to be exploited on commercial basis for export strategies. Besides, there are areas of investment such as teak processing, electricity generation, cutting and polishing of precious stones, cutting and polishing of granites, safety matches making units, milling of pulses and oil seeds, etc. Trade and commodities exchange market may be established both at Moreh and Imphal. One at Moreh would provide commodities of daily use while the one at Imphal may be stocked with high value items like engineering items, electrical equipments, metals and minerals, jewellery etc.

Quantum of trade in manufacture commodities can increase in the near future. Therefore, it is necessary to develop an all-round productive industrial base at Imphal. The centre at Moreh may be used for processing, storage and warehousing purposes while the one at Imphal may be organized as manufacturing complex.

Success of China-Myanmar and Thailand-Myanmar border trade is primarily due to free and fast movement of cargo by road. The same will be true for the Indo-Myanmar border trade if substantial productivity in the North East along with standard road network is established. Therefore, National Highways like NH 39 and NH 53 need to be converted into 6 lane highways. In this regard upgradation of the Second World War road, the 160 km Tamu-Kaewa-Kalemyo Road, will mean easier movement of people and commodities. This road, which links the town of Moreh to central Myanmar will soon link Mandalay, Myanmar's second largest city, and with Yangon and even Bangkok. Improvement of NH 39 as part of the Asian Highway Project, which identifies 55 Asian routes among 32-member countries, must be taken into account as it is related to facilitation of land transport and mobility.

Conclusion

Manipur State is immensely dependent upon imported commodities. The degree of interaction shown by the flow of commodity greatly influences the economic base of the State. The transport linkage developed by the inflow of commodity is stronger than the linkage established by the outflow. Hence, lopsided pattern of flow between the State and the rest of the country significantly underlines the nature of underdevelopment witnessed so far. Manipur's socio-economic condition can improve only when there is a strong transport linkage between the State and the rest of the country in the backdrop of substantial production in the State. In this respect, adoption of NH 39 as part of the Asian Highway is crucial.

REFERENCES

Dun, E.W. 1886. *Gazetteer of Manipur*. Calcutta: Superintendent of Government Printing.

Hall, D.G.E. 1955. *A History of Southeast Asia*. London: Macmillan.

Hervey, G.E. 1993. *An Outline History of Burma*. London: Longmans.

Indian Institute of Foreign Trade. 1965. *Report on Promotion of Border Trade between India and Myanmar*, New Delhi.

Pemberton, P.B. 1966. *Report on the Eastern Frontier of British India*. Gauhati: Department of History and Antiquarian Studies.

Singh, R.K. Jhalajit. 1965. *A Short History of Manipur*. Imphal: R.K. Book Store.

Singh, R.P. 1982. *Geography of Manipur*. New Delhi: National Book Trust.

Singh, Yambem Imo. 2006. Spatial interaction between Manipur and the rest of the country: An evaluation. Ph.D. thesis, Manipur University, Manipur.

Sharma, B. Lalhari. 1989. *Transport Network in Manipur: A Geographical Analysis*. Ph.D. Thesis, Manipur University, Manipur.

Sharma, Nandalal. 1960. *Meitrabak*. Imphal: O.K. Book Store.

Index